The Little Man with Tremendous Power!

MIGHTY ATOM

(Joseph Greenstein, of Houston, Texas)

IN PERSON

MARVEL OF THE AGE!

THE GIANT OF TODAY!

LITTLE, BUT OH MY!

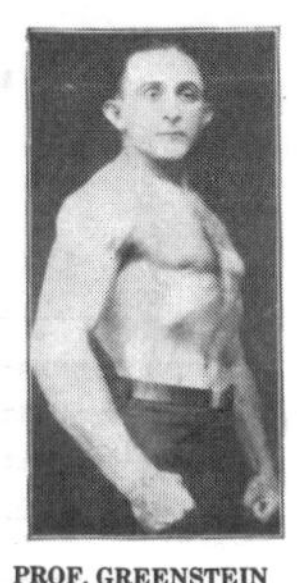

PROF. GREENSTEIN

The Modern HERCULES
The Young SANDOW
The Ancient SAMSON
ALL IN ONE

WORLD'S STRONGEST MAN

YES, Smallest Strong Man in the Universe!

HIS PRIVATE LIFE:—Age, 47 years, Weighs only 148 lbs., 5 ft. tall. Has wife and 10 healthy children on vegetable diet mainly.

זיין פּריוואַטער לעבען וועגט נאָר 148 פונט, 5 פוס הויך. זיין ווייב האָט 10 קינדער
אלע 100 פּראָצענט געזונט הויפּטזעכליך אויף א דיעטע פון וועדזשעטייבעלס.

FEATURED OVER R. K. O. VAUDEVILLE CIRCUIT AT $3000 WEEKLY BEFORE DEPRESSION

געשפּילט איבער די אר קעי אָ וואָדעוויל סירקוט קריגענדיג דריי טויזענד דאָלאר
א וואָך בעפאָר די דעפּרעסיע.

Has Received Keys From The Mayors of 19 Cities! Special Honor Guest Everywhere!
Was Presented with 11 Medals from World's Highest Officials!

ער האָט עלף מעדאלען פון די געקרוינטע קעפּ פון אייראָפּא.

He will give EDUCATIONAL HEALTH LECTURE before act.

גיט עדיוקיישיאָנאל העלט לעקטשורס בעפאָר ער ציינט זיינע וואונדער.

JUST A FEW OF THE "UNBELIEVABLE" FEATS YOUR EYES WILL SEE!

נאָר עטלעכע פון די „אומגלויבליכע" זאכען װאָס אייערע אויגען װעלען זעהן.

(1) Bites big steel nails in half with his teeth.

1. בייסט גרויסע שטאָלענע טשוועקעס אין העלפט מיט זיינע ציינער.

(2) Snaps dog chains wrapped around his arm muscles.

2. פּלאצט הינטישע קייטען ארומגעבונדען ארום זיינע אָרעמס מוסקלען.

(3) Bursts heavy chairs by expanding his chest.

3. שפּאלט שווערע קייטען דורך אויסשפּרייטען זיין ברוסט.

(4) Bends heavy iron rods with his fingers.

4. בויגט שווערע אייזענס מיט זיינע פינגער.

(5) Makes finger rings out of 20 penny nails.

5. מאכט פינגער רינגען פון 20 פעני נעגעל.

(6) Twists 2-inch steel bars around his arm as bracelets.

6. דרעהט ארום 2-אינטשיקע שטאָלענע בארס ארום זיין אָרעם אלס בראסלעטען.

(7) Plays with 200 & 300 lb. dumb-bells as baby plays with toys.

7. שפּילט זיך מיט 200 און 300 פונטיגע דאָמבעלס אזוי ווי א בייבי שפּילט זיך מיט שפּילצייג.

(8) Bends heavy iron bars by pulling on his hair with iron comb.

8. בויגט שווערע אייזערנע פּאיפּס דורך שלעפּען אויף זיינע האָר מיט א אייזערנע קאָם.

(9) Supports big horse, cow, and 4 men on his chest while he lies on bed of iron nails.

9. האלט 4 מענער אויף זיין ברוסט און וואַרפט זיי ארום אזוי ווי א פּילקע.

(10) Supports 4 men on his chest and bounces them as a rubber ball.

10. האלט אָפּ א עראָפּלאן וואָס לויפט 150 מייל א שטונדע מיט זיינע האָר.

(11) Stops airplane at 150 miles speed with his hair.

11. שלעפּט א 20-טוניגען פייער טראָק ארום א בלאָק מיט 20 מענער.

(12) Pulls 20-ton Fire Truck around block with 20 men.

12. האקט אריין 20 פעני נעגעל אין האָלץ מיט איין קלאַפּ פון זיין נאקעטען פויסט.

(13) Drives 20 penny spikes through 2-inch wooden plank with just one stroke of his bare fist.

13. בויגט אייזערנע שטאנגען אויף זיין נאָז אויסהאלטענדיג פיער מענשען.

(14) Bends iron bars on his nose while suspending four people.

14. גלייכט אויס פּאָדקאָוועס מיט זיינע הענט און צוברעכט זיי.

(15) Straightens horse shoes with his hands and breaks them.

15. לאכט זיך אויס פון דערשטיקט ווערען פון צוויי מענער דריקענדיג זיין גאָרגעל מיט א אייזען צום וואנט.

(16) Defies strangulation by 2 men pressing on his throat with rod against wall.

16. ווינט א מערינגאָראונד פון 6 מענשען הייגענדיג אָן זיינע האָר.

(17) Swings Merry-Go-Round of 6 people hanging to his hair.

17. האלט אָפּ צוויי האַי-פּאור אָטאָמאָבילען וועלכע רייסען זיך אוועקצולויפען ווען ער האלט זיי אָן מיט שטריק.

(18) Defies two high-power automobiles to break away while he holds both with rope.

18. האלט א גרויסען פערד, א קוה און 4 מענער אויף זיין ברוסט ליגענדיג אויף א בעט פון אייזערנע נעגעל.

Some U. S. A. Press Comments About the Mighty Atom:

"The modern Samson"—"Believe It or Not" Ripley, N. Y. American, July 1, 1931; "Does the seemingly impossible"—N. Y. News; "Has tremendous strength"—N. Y. Times; "Young Hercules" — Chicago Tribune; "Resembles Goliath of the Bible"—Boston Post; "Positively 'little, but oh my!'"—San Francisco Express; "Marvel of the Century" — Philadelphia Despatch; "Strongest of the present day"—Cleveland Plain Dealer; "The iron man of this age" —Washington News; "Must be seen to be believed" — Los Angeles Examiner; "Such Herculean stunts from so little a man" — Seattle Oregonian; "Texas produced world's strongest man" (He is from Houston) — Dallas Herald; "Has muscles of iron" — St. Louis Chronicle; "Seems inhuman endurance"—Milwaukee Sentinel; "Supernatural resistance" — New Orleans Item; "What right living can do" — Kansas City Star; "Miracles come to life" — St. Paul (Minn.) Picayune; "Has hair of steel"—Maine Advertiser; "Body seems of stone"—Detroit Press "Nothing like this modern giant ever seen"—Cincinnati Recorder — & 1000 MORE!

HE HAS 11 MEDALS FROM THE CROWN HEADS OF EUROPE!

SEE OTHER SIDE — FEAST YOUR EYES ON THE MOST GLORIOUS SIGHT of your LIFE!

THE MIGHTY ATOM

THE MIGHTY ATOM

THE LIFE AND TIMES OF JOSEPH L. GREENSTEIN

ED SPIELMAN

THE VIKING PRESS — NEW YORK

First published in 1979 by The Viking Press
625 Madison Avenue, New York, N.Y. 10022
Published simultaneously in Canada by
Penguin Books Canada Limited

Library of Congress Cataloging in Publication Data
Spielman, Ed. The mighty atom.
1. Greenstein, Joseph L.
2. Strong men—United States—Biography. I. Title.
CT9997.G73S65 791.3'5'0924 [B] 79-12911
ISBN 0-670-47564-5

Printed in the United States of America
Set in Linotron Century Schoolbook

The endpapers, frontispiece, and photographs on pages
x, 17, 23, 24, 59, 67, 75, 92, 95, 106, 109, 125, 135,
165, 175, 178, 190, 210, 213, 221, 230, and 233
are from the author's collection.

This book is affectionately dedicated
to my mother, Harriet Moss,
for her years of love and sacrifice,
and to my wife, Bonnie,
without whom it would not
have come to fruition.

I wish to thank the following people for their kind cooperation and assistance: The Greenstein family, Cathy Bardram, Randall G. Bassett, Victor Boff, Mrs. William L. Champion, George Dillman, Mollie Cohen Greenstein, John Grimek, Naraki Hara, Lew Horn, Jesse Kahn, Esq., Charles Kornhaber, R. E. Liebmann, Maria del Socorro S. Martinez, Theron K. Rinehart, Irving Shapiro, Abe Shenkman, Al Spielman, Sam Spielman, and Andrew Yelaney.

Special thanks to: my brother, Howard Spielman; Lawrence "Slim the Hammer Man" Farman; my literary agent, Peter Lampack; my friend and colleague of long standing, Howard D. Friedlander; and Alan Williams of The Viking Press, for his insight and invaluable assistance.

PROLOGUE

In the days of the amusement palaces, when Coney Island was the playground of the world, the crowds stood twenty deep several times a day to watch the last of the great strongmen perform the impossible. My family were often among the spellbound. My grandparents Jacob and Jennie Shapiro had made the transition from poverty on New York's Lower East Side to Coney Island, an enclave of working-man's luxury where hot and cold fresh and salt water was piped into the apartments. On Sundays Jacob, a gentle bear of a man, with his daughter Harriet on one hand and son Irving on the other, stood transfixed as the Mighty Atom performed his wonders. In earliest childhood I had been told by my family about this man; later, I believed him to be an exaggeration long lost in the dim past.

I first met the Mighty Atom at a martial arts show in Madison Square Garden. I was one of the featured guests, having created the *Kung Fu* television series which sparked the Kung Fu craze of the 1970s. I took my bow with David Carradine, some exploitation actors, and sundry exhibitionists before an audience of eighteen thousand devotees of the Bruce Lee cult. Various martial arts were demonstrated: man nearly cuts his buddy's genitalia off with an imitation samurai sword. Applause. Two-hundred-twenty-pound man beats a 120-pound woman unconscious. More applause.

Then the old man appeared and walked up past the ring apron. His white hair flowed down to his shoulders; his brown leather tunic bore a gold Star of David. He looked like a living Maccabee. He was a short man, only 5 feet 4½ inches, but broad; in his strongman's costume he was a miniature Samson.

The crowd watched in silence as he bent horseshoes with his bare hands and bars of steel across the bridge of his nose, exploded chains with the expansion of his chest, and drove spikes through planks of metal-covered wood with nothing more than the power of his palm. Then the crowd came to its feet, and with good reason. This man was eighty-two years old. A half century after my grandfather had first seen him, I sat ringside, watching the strongman. I could hardly believe that Joseph "the Mighty Atom" Greenstein was still alive. As if deposited by a time machine, he stood in the center ring of Madison Square Garden and seemed to transcend age and time.

It was not sheer animal strength that awed the crowd, but something greater—an inner power. One's eyes were drawn to him; he was ancient, yet vitally alive, a compact boulder of granite that pounding water had washed and smoothed. Somehow, it was his mind that held you, not his body. This was no artifact of dumb muscle, but a man whose intelligence expressed itself in every line of his face.

I decided to visit him backstage, although I harbored no small suspicion. At thirty, I was successful by the standards I had set for myself; but broken promises (always accompanied by porcelain smiles and firm tan handshakes) had made me cynical. When a Hollywood screenwriter has had his heart and cinematic vision vandalized a dozen times but is still resolute, he is called a professional. I found myself disaffected with an industry that twists wonderful screenplays into mediocre films. How significant to me was the opening scene of the film *Sunset Boulevard*, in which the screenwriter is found floating face down in the pool.

And so, in meeting the Mighty Atom, I maintained some reticence, afraid I would discover that he, too, was just another phony spellbinder, another product of show-biz hype. I could not face another disillusionment, confront the paling of another romantic image beside the reality.

I introduced myself. He was gracious, very soft-spoken, and his speech was flavored with the lyrical Yiddish accent of my grandparents. He invited me to his home. When we shook hands, he did so gingerly, as if afraid to crush mine. He had a good left eye; the right was blind and milky white. But that one good eye of crystal blue had the power of a Svengali. A life-force emanated from him, as if he had found the secret, the way to deny old age and death.

Some days later I went to see him. The neighborhood surrounding his brick row house on East Ninety-sixth Street and Rutland Road in Brooklyn was once the heart of a prosperous middle-class community. Now, glass was strewn in the streets and groups of stray dogs ran wild. Those of the old neighborhood had fled—all but one.

A faded sign was stuck in the patch of front yard:

INTERNATIONAL HEALTH PRODUCTS
MIGHTY ATOM, PROP.

He came to the door; his shoulder-length hair was pinned under his skull cap, the powerful chest and wrestler's arms concealed by a shirt and vest. He sported several lapel pins—one a memento presented by the Israel Defense Forces, another a prize for poetry.

He lived in an apartment at the rear of the multifamily house, with no woman to care for him. A wall plaque with the Hebrew word *Shalom* hung over a kitchen table cluttered with jars of pens, sheaves of Yiddish poems, and mail orders for Mighty Atom products from the one-man health company he had operated for fifty years. His vest matched neither his pants nor the jacket that was draped over the chair. He kept a pair of snow tires in the kitchen, an antique washboard and a fifty-year-old can opener were on the sink, Mighty Atom soap and liniment were in the bathroom.

He intended to sell the house and leave. We went down into the cellar to review the artifacts of a lifetime. Late at night, our only illumination coming from a bare bulb that hung over our heads, we were a pair of archeologists out to excavate his past glories. Rusted lengths of metal hung on an old display board, twisted into corkscrews as if by some uncontrollable force of nature, contorted into curliques like coiled auto springs. He could name the place and year for each object. He smiled when I found the key to the city of Saratoga Springs under a half inch of dust in a remote drawer; it was, he said, one of twenty such keys, and most of the others were destroyed in the great Coney Island fire. Bags of old fan mail sat by vaudeville posters and photos of the Atom in action.

The old man paused to reflect over a beaded bag that had been a favorite of his wife's. Knowing full well that it would end up in the garbage when the next owner arrived, he put it in the "save" box, although he understood that he couldn't take it when he moved to a room in his daughter's house. A rare assortment of stuff: a .32 Smith and Wesson revolver that he had taken from a hooligan in the South of the 1930s, ancient Hebrew books, and others on Jewish history. We finished our treasure hunt after three hours.

The Atom and I talked late into the night. It was a dangerous neighborhood and over my protests he escorted me to my car. Despite my embarrassment I knew he was right: he was the Mighty Atom still.

I came to see him again and again, always with affectionate skepticism—waiting for self-aggrandizement, which never came—and never quite knowing where it would all lead. Days became years. I absorbed the myriad details of his life, though he often glossed over the most fantastic occurrences or feats with a mere nod or wave of his hand. "Is that true, Joe?" I would ask, and inevitably he would smile and, without seeking approbation, reveal printed documentation. He wished to relate his life to me because he knew that his years were growing short, and believed that if it were not all recorded somehow, it would be as if he had never lived it.

This is his story, as he told it to me, buttressed by decades of newspaper and magazine articles, personal interviews with eyewitnesses, and my own observations and research. I have filled in a line of dialogue here and there, a missing conversation. Of his life, times, and feats I have invented nothing. The story is true.

CONTENTS

He has made a marvelous fight in this
world, in all the ages; and has done
it with his hands tied behind him. . . .
The Jew . . . is now what he always was,
exhibiting no decadence, no infirmities
of age, no weakening of his parts, no
slowing of his energies, no dulling of
his alert and aggressive mind. All
things are mortal but the Jew, all other
forces pass, but he remains. What is
the secret of his immortality?

—MARK TWAIN

There is no such thing as a little man, and
nothing is impossible.

—JOSEPH L. GREENSTEIN

THE MIGHTY ATOM

YOSSELLE

The heat was oppressive on July 15, 1893, the third day of the month of Tammuz in the Hebrew calendar. Chaya Greenstein, breathing heavily, carried water up Malorotczik Street in the poorest section of Suvalk, Poland, a garrison town near the German border. The woman, in the sixth month of her eighth pregnancy, stumbled and fell. Labor began almost immediately, and her baby was born soon after, a tiny, wizened thing weighing only three and a half pounds. Its eyes remained closed; only the faintest whisper of breath issued from its lips. She placed it in a small box, on a layer of cotton she had saturated with poppy-seed oil to protect the skin, and nursed it with an eye dropper.

In those days premature babies rarely survived, and when the child, which had been named Yosselle Leib (Joseph Lewis), endured for a week, news of the event spread. On the ninth day of Yosselle Greenstein's life, his father Herschel, a rabbi who was obliged to eke out a living at unscholarly labor as a tanner, looked up from his book to see three well-dressed gentlemen trudging through the ankle-deep mud toward the house. They were three doctors, Anievitch, Bokanofsky, and Nathanson, who had come to view the unusual infant. Herschel let them in and stood nearby as they gathered around the child. There was nothing they could do, they said. They inspected the biological curiosity for some time, and before leaving diplomatically told the Greensteins that if the baby should expire they would pay a sum of money for the body, for use in their medical studies.

But the child clung to life. He completed his development slowly, outside the womb. The eyes opened, the hair and nails grew. He struggled for breath, too weak to offer a healthy howl. At the age of normal birth, nine months and eight days the child was circumcised according to Jewish law.

The Greensteins were the poorest family in Suvalk. They had little food—seldom more than potatoes found in the fields or old bread. When Yosselle was five his father died, and despite Chaya's struggles to provide for her children, life became more bitter. Even in winter Yosselle had no shoes; he walked to school with his feet wrapped in rags. He was pale and sickly, an asthmatic like his father. But as the youngest member of the family he was especially cherished. He had an endearing manner and clear blue eyes that held one's gaze, as if some inner force were transcending the frailty of the body.

The winter after his fourteenth birthday, his mother took him to the doctors who had visited him when he was a baby, and after a lengthy examination, they offered a medical opinion.

"Mrs. Greenstein," said Bokanofsky, "your husband passed on at thirty-nine of respiratory illness. It seems your son has inherited his father's infirmity."

"This asthmatic condition," said Nathanson, "appears congenital."

"We feel you should know," Anievitch added quietly, "we don't believe this boy will see his eighteenth birthday."

In the next room Yosselle heard the doctors predict his death.

On the way home, the woman attempted to be cheerful, biting back her tears. They passed a traveling tent show set up not far from the road. The boy looked up to see a man with steely eyes and a black waxed handlebar mustache. He was not so tall as wide, and his physique resembled chiseled rock more than muscle as he stood with arms raised in a strongman's pose. It was a circus poster emblazoned with pictures of wrestlers who were part of the show.

On the placard was the name of the star attraction, "Champion Volanko."

Yosselle, shivering in his ill-fitting overcoat and coughing into his handkerchief, stood captivated by the poster. Nearby, a man named Durov was putting his trained dogs through their paces. Durov was a tease for the crowd, a promise of wonders to come. His was a fine entertainment, and he had performed before the Czar.

"It's a show, Ma. Can I stay?" he asked his mother.

"A while, Yosselle," she said. "But come home soon. We'll sit by the fire."

The boy stood in the crowd as Durov's animals entertained, and when he had finished, those with money began to file in. Admission was the equivalent of fifteen cents. Yosselle turned his pocket inside out; there was nothing at its bottom but a hole. He noticed some Polish boys sneaking in under the tent, so he picked out his own remote corner of the tent to try his luck. As he attempted to make his way under, he was seized roughly by the neck and pulled inside. He found himself in a dark corner of the tent with a large man looming above him.

"Dirty little Jew," the man spat at him. "I'll kill you."

The blow of a fist knocked the boy down, and heavy boots exploded into his head and sides. This was no mere beating; the man indeed intended to kill him.

Sprawled on the sawdust, the boy had the presence of mind to feign death, not moving even as the last of several finishing kicks slammed into him. Then, at last, he heard the sound of footsteps leaving the tent.

His eyes opened slowly. He wiped the caked dirt from his face and his hand came away wet with blood. He was too weak to stand. Through glazed eyes, he saw a flickering light across the darkened rear of the arena. He began to creep toward it. Between him and the beacon, circus horses were tethered, but he crawled under their bellies, ignoring the occasional impatient hoof that struck the ground near his face. Inch by inch, he pulled himself toward the light.

A simple kerosene lamp had beckoned him, shining from within a wagon standing beside the tent, a crude dressing room. In a corner was a table with a black cloth draped over it. With the last of his strength, he dragged himself in, hid beneath the table, and lay there in shock, huddled in that hiding place for what may have been hours or minutes. At the sound of heavy footsteps, his eyes widened. With a start, he peeked out of a small rent in the fabric, as a man entered the wagon, a gargantuan figure with granitelike features, a black, waxed mustache, and unwavering blue eyes. He had a barrel chest, arms like the limbs of an oak, and wore the knee-length boots of a Don Cossack. As he brushed past the table, the big man sensed the human presence beneath it. "What devil is this?" he exclaimed in loud Russian and ripped the cloth away. In terror, the boy shielded himself from further blows.

The sight of the bloody little wraith beneath the table took the big man aback. He stared at the boy, then knelt and lifted him without effort. The voice was oddly gentle. "It's all right, son. Don't be afraid." It was the show's star wrestler, the likeness on the poster—Champion Volanko.

He sat Yosselle on a chair, took down a box of bandages and ointments, and tended to his wounds.

"What's your name, son?"

"Greenstein . . . Yosselle Greenstein."

"Who did this to you?" the wrestler demanded, but the boy could not respond.

"Wait here." Abrubtly Volanko rose and left the tent.

Bewildered, Yosselle looked around the room. There were fancy costumes and polished boots, a shelf of books, and spartan but homey furnishings. His observations were interrupted by Volanko's booming voice: "Boy, come out!" Outside the dressing room nineteen men were lined up in the corner of the tent where the beating had taken place. "Show me the man who beat you," Volanko demanded.

"I can't . . ." Shielding himself from the blows in the half-darkness, he had not see the man's face clearly. Volanko rubbed his chin in disappointment.

"But he spoke. . . . I heard him," Yosselle said.

"You would know his voice?"

Yosselle nodded with certainty; he had a musical ear and could remember voices well.

"What did he say?"

"He said, 'Dirty little Jew . . . I'll kill you,' " Yosselle replied.

The men stood nervously as Volanko ordered each to repeat the invective. One by one they did so. One by one the boy shook his head. This went on until the seventeenth man, third from the right, a burly peasant in his twenties. Volanko ordered: "You. Say it." As the man spoke, the boy recoiled, his very reaction an accusation. Volanko knew. This was the man.

As Volanko advanced on him, the peasant protested that he had indeed beaten the boy. "But he was trying to sneak in. It was my duty . . ." The words were barely out of his mouth when Volanko was upon him. Seizing him by the collar, the wrestler lashed out with a single blow, crunching the man's nose flat on his face and hurling him aside like a rag doll. So violent had been Volanko's attack that he was left standing with the now senseless man's torn lapels in his hand.

Back in the dressing room, the boy wondered to himself why this man, a Jew-hating Don Cossack, should help him. But Yosselle did not ask.

A few moments later, Volanko put on a red wool robe, slippers, and rimless reading glasses, then poured the boy a cup of tea and sat down in his armchair. Yosselle coughed, and the wrestler looked at him with a knowing expression.

"You've seen a doctor?"

"Three of them."

"And?"

"They say I will die soon."

"Doctors . . . bah . . . bloodsuckers! They tell that to everybody." He smiled. "What if I told you that I, Volanko, was once more sickly than you?"

"Is it true?"

"Is this the face of a liar?" Volanko waved a finger for emphasis. "The greatest athletes have grown from the weak and infirm." He paused for a moment. "Do you want to die?"

"No."

"Then don't," Volanko answered with certainty.

"You could cure me?"

"No. Only you could do that. I might show you the means." Abruptly, the wrestler's enthusiasm vanished. He shrugged. "Pity, the circus leaves tonight."

The boy's expression fell. He coughed again. Volanko studied one of his perfectly polished boots with displeasure.

"Tsk, tsk. Poor shine. If only I had a valet."

"*I* could be your valet," the boy said, brightening.

"Now, why didn't I think of that?" the wrestler replied. "Do you have a family?"

"Yes."

"A mother?"

"Yes."

"Go home and get her permission. Be quick. We leave when the sun is down."

Yosselle knew that his anxious mother would never let him go, and so he decided not to go home at all. He stayed in the village for a while and gathered some rags which he stuffed into his handkerchief so that he would give Volanko the impression that he had collected his belongings. He returned just as the train of carved, brightly painted wagons was pulling out. The wrestler pulled him aboard, and from the open door of Volanko's van, he watched Suvalk, the only world he had ever known, recede into the distance.

Yosselle awakened from sleep in his cot in a corner of the wagon. He pulled the covers up under his chin against the cold, rubbed a circle of frost from under the isinglass window, and peered out.

In the dim light before morning, he saw a figure ambling through the snow. It was Volanko. The big man stopped several yards from the wagon, removed his robe, and hung it on a tree. Standing bare-chested and wearing only his long drawers, he reached down and scooped up handfuls of snow and smeared them on his face and chest and under his arms. His deep breaths became vapors clouding the air as he reveled in his snow bath like a polar bear, uttering invigorated oohs and aahs.

As Volanko finished with his morning regimen and headed back toward the wagon, Yosselle nestled under his blanket as if asleep. Volanko walked straight away to his cot and shook him.

"Get up, my friend. Enough sleep."

"What time is it?" The boy clung to his blankets.

"Very late. It'll be dawn in an hour. This is the best time."

"For what?"

"For everything," the wrestler said.

Out they went, a bundled Yosselle still hunching his shoulders in the cold. "I'm hungry," he protested.

"Later. Now is the time to breathe. Go ahead."

The boy inhaled thinly, exhaled an asthmatic wheeze, then coughed.

"Tsk, tsk . . . terrible. You cheat yourself, little friend. The air is free. Look, like this." Volanko sucked in air through his nose while bringing his hands together over his head. Then he exhaled through his mouth and pumped his arms, working his lungs like a bellows. "Breathing is life, Yosselle. Without air, the fire dies. Now try again."

The boy tried to imitate him, and Volanko closed one eye with a pained expression as his emaciated valet puffed away. "No, no, you'll faint if you do that. Slowly . . . relax. . . ." The boy tried again. "Yes, more like that."

"How long do I have to do this?" the boy asked gasping.

"For the rest of your life," Volanko replied.

Yosselle longed to see his mother, his sisters and brothers. He knew they were in mourning for him. It was better this way, for them to grieve and forget him. He would die, or return home a new man. He would not write home.

That morning Yosselle stood anxiously in line with the circus hands to get breakfast, a ladle of white mush. The aroma rose to his nostrils; in all his life he had never smelled anything so good. He sat on a barrel and was about to dig in, when a hand snatched the treasure from under his nose. Volanko stood angrily over him holding the steaming plate. With a crook of his finger, he motioned for the boy to follow. "Mr. Volanko, please . . . I'm hungry," Yosselle begged him tagging along almost in tears. The wrestler maintained a stony silence.

"Please," Yosselle cried, "I'll share it!"

"This is not food." He spat in the plate. "It's garbage."

Confused, the boy pointed to the others. "They eat it."

"What do *they* know?" Volanko retorted, scowling.

They passed the cage of Durov's trained dogs. "A man who eats like a dog"—Volanko slid the plate under the bars—"lives like a dog." The animals gobbled the food down— "and dies like a dog." The big man then marched Yosselle back to his wagon, where a pot and kettle simmered on the wood stove. Volanko dished up a bowl of steaming grainy brown cereal and offered it to the boy, along with a cup of herb tea. "From now on, you eat only what I give you, understand?"

The wrestler produced his pocket watch and chain. Flipping open the back of the timepiece, he revealed a ticking maze of tiny wheels and springs. "Look inside, Yosselle. Look how fine . . ." He tapped the boy's chest with a thick finger. "Inside your body is a machine a million times more fine and more important. What if I were to grind a handful of dirt into this timepiece?" He frowned. "Disgusting, no? Terrible. Like this watch, you don't put in your body what is unclean or bad for it."

The boy shoveled the hearty breakfast into his mouth, digesting Volanko's words with his food.

As they traveled in the weeks to come, the locomotive heaved black smoke above the snowy landscape and hauled its cargo of circus wagons atop flatbed cars. The circus moved toward warmer quarters, setting up in town after town across Eastern Europe, and Champion Volanko's bootblack, valet, and aide-de-camp observed, listened, and learned.

"Boy, I am not your mother," Volanko admonished him. "I will not pat your head and sing you to sleep. Such kindness you can do without. There is an illness in you, Yosselle. Either you will kill it . . . or it will kill you. Now which will it be?" He arched an eyebrow.

"I will kill it!" The boy laughed with anticipation.

"For sure?"

"For sure."

"You will?" Volanko taunted.

"I will!" The boy coughed and gritted his teeth with resolution.

"Breathe!" the big man commanded.

Yosselle inhaled the early morning air; it stung his lungs and he could hear Volanko's voice behind him. "Set your mind, boy. Refuse to be weak, refuse to be sick, refuse to die. Think that you're strong and you are."

By inches over the next weeks the wracking cough loosened its grip on the boy, and his pallor was replaced by the faintest hint of pink. Volanko congratulated neither of them, but soon he interrupted the morning routine, presenting Yosselle with a pair of wooden buckets.

"Yosselle, as you breathe, raise these over your head."

The buckets were sturdy. "I don't know how many times I can do it," he said.

"Could you raise them once?"

The boy nodded, and lifted them awkwardly.

"One more," Volanko demanded. When that was done, "One more . . . only one more." When the boy could no longer go on, he had raised the pails ten times.

Volanko sat him down to rest. "If I had told you to raise them ten times, you probably couldn't have done it," Volanko said, assuming his philosophical expression. "Yosselle, even a task that is terrifying you can do, if you don't think too much. Look at the whole of it, and you'll lie down exhausted before you start. Instead, do only the small job in front of you. Only that . . . one brick . . . one step . . . only that. Think no further. Then do one more . . . one more . . . one more . . . only what is in front of your face. Before you know it, you'll have done it all—without fear or concern. Time will work for you. There is no other way."

The next morning Volanko once again handed Yosselle the buckets, and tossed a handful of sand into each one. The boy raised the buckets, now infinitesimally heavier, over his head. "Remember, Yosselle," said Volanko "Time will work for you."

Yosselle repaid Volanko with work. He kept their quarters and the wrestler's wardrobe and boots immaculate, and performed his tasks before being asked. In tidying up the wagon, he picked up Volanko's reading glasses, opened a cupboard to put them away, and abrubtly stopped. Inside was an old prayer book with a gold Star of David on it. Astonished, Yosselle picked it up and looked at the worn pages of Hebrew script. Hearing a sound behind him, he turned to find Volanko standing in the doorway. A long moment passed before the big man spoke. "It was my grandfather's, my father's, now mine," Volanko said in Yiddish, "But Jews do not wrestle with the circus."

Volanko paused and sat, indicating that Yosselle should do the same. "We live in a strange world, Yosselle. While many accept our laws, revere our prophets, benefit from our talents, and give their children our names, they get upset by our presence." The boy nodded his understanding.

"I never met anyone like you."

Volanko's brow furrowed. "By the time I was your age, I had survived three pogroms. So, like you, I ran away to be free."

"And now you are Champion Volanko. . . ."

A shadow of guilt swept Volanko's face. "A man who denied himself and his people . . ." He spoke emotionally, with a profound sadness. "Learn from my life, Yosselle. Don't pay the price I paid. It's hard to be a Jew, hard to be a man . . . but try. And never forget who you are."

As the most skilled of the circus wrestlers, it was Volanko's responsibility to take on local challengers wherever the circus happened to be. These bouts, popular with audience and circus folk alike, increased attendance. In preparation, Volanko often sparred with other circus wrestlers in the Graeco-Roman style. But this morning Volanko passed them by, his valet at his heels, explaining, "I have an opponent far more important."

Moments later, outside the circus tent, Yosselle stood trying to subdue Volanko with an arm lock, a ludicrous mismatch.

"Hands higher. Feet apart. Good . . . now throw me," Volanko demanded.

"I can't," the boy protested.

"*You can.*"

Yosselle exhausted himself trying.

"Yosselle," Volanko instructed, "before you begin a task, you must first succeed . . . up here." He raised a finger to his head. "Mind and spirit . . . then action. Like a bullet fired from a gun. Once it begins, it's already done."

Volanko paced for a moment with his eyes on the ground. "Look"—he pointed down, then got on all fours—"an ant, a speck of a creature, carrying a burden many times his own size." The boy knelt for a closer look.

"How does he do it so easily?" Yosselle asked.

"He's an insect." Volanko smiled wryly. "He doesn't know he can't."

As the show traveled east from Poland into Russia, Yosselle mingled with spangled equestrians, dancing bears, men and women who risked death in the air—a netherworld of those who earned their bread by skill and daring. But Volanko alone was his teacher, imparting his knowledge and training. The mental, physical, and spiritual were treated as parts of the whole, for Volanko taught him if one was disturbed, the others would suffer.

Yosselle "killed" his cough, and the possibility of becoming a wrestler like Volanko, at first a daydream, now became real.

Some months had passed, and it was evening, when the wrestler put down his reading and studied his valet for a moment. By now the boy's uncertain heart beat had grown much stronger, and he had gained healthy skin color and fifteen pounds of muscle. Volanko leveled a finger at him. "Yosselle . . . outside."

By the light of the moon, the wrestler began to show him another breathing form. "Extend the stomach while taking air in, contract while exhaling. . . ." At the end of his long day, the boy was bored and a bit detached. "What can this do for me?" he asked, obviously convinced he had mastered the technique. Volanko produced a string and tied it around the boy's chest. "Break it," he ordered, "and don't use your hands."

Yosselle attempted to split it with his chest, but his expansion was insufficient, and the thread only cut into his skin. "Keep trying," Volanko said and went back inside. For half an hour Yosselle twirled about trying to overcome the bit of string. When Volanko came out of the wagon again, he was carrying a length of chain. "Yosselle, you're a good boy. And you learn quickly. But don't let vanity close your mind."

The wrestler swung the chain around his back and held it across his chest. "Master low and high breathing, and you control the brain, the heart, and the other organs. All bodily power comes from the breath."

The wrestler took in air, extended his stomach as if he had swallowed a balloon, and fastened the chain across his chest with a steel link.

"This is what I've been trying to teach you," he said, then rapidly contracted the stomach and shifted the air up to his chest. He strained for a moment. Then the chain exploded.

Volanko's bootblack stared at the ground.

"It's all right, Yosselle," the wrestler said. "When I was your age, I too was very smart. I'm much dumber now." He winked.

From Kharkov in the Ukraine, the circus made its way across Eastern Europe past the Aral Sea to Tashkent. Late in 1908, they arrived on the unbroken fields of Poona, India, a mountain city which was the former capital of the state of Maharashtra, near Bombay. Volanko stood beside the tiger's cage with his hand thrust through the bars, patting the neck of a grown Bengal.

"Ah, Tascha, you know you're home. India. The scent of your mother is on the wind."

Volanko took Yosselle on the several hours' journey to Bombay, where they made their way down boulevards lined with grandly ornamental buildings. Fierce Punjabi mounted police, men with crimson turbans, black eyes and beards, and bright uniforms stood watch as sloe-eyed women in delicate silk saris paraded, some with gold ornaments in their noses. The city was a riot of noise and color, but in the native quarters and bazaars Yosselle found the habits of the common man unseemly; they chewed betel nut and expectorated awful red stuff onto the sidewalk. He was repelled by the flocks of beggars, some armless or otherwise mutilated, who pursued them, crying out for charity. At last, he and Volanko arrived at a place which was crowded with a chattering throng; they were the only Europeans among them. It was a wrestling ring. For centuries wrestling was the national sport of India, and it was treated with some formality.

In the center of a loose earthen pit, two massive Indian wrestlers stood naked except for breech cloths, their black mustaches preened and heads shaven. Each raised a weighty gold scepter-mace, his badge of championship. This formality concluded, the two men faced each other in the open pit. The crowd took up a chant, calling the name "Gama" again and again as the match began. In a matter of seconds, the smaller man upended his opponent and slammed him to the earth as a roar went up. A gross mismatch. The victor was Volanko's friend, "Gama, the Lion of the Punjab," who would in years to come be recognized by Western sports historians as the greatest wrestler of the twentieth century.

After the match, Volanko, Gama, and the boy sat in the shade as the two wrestlers talked idly and all shared from a large basket of fruit, a gift to Champion Gama who was a strict vegetarian. With appropriate deference, the boy addressed him.

"Gama, may I ask you something?"

"Ask."

"Your opponent was very big . . ."

". . . and yet I threw him like a baby."

"How?" Yosselle asked in wonderment.

"It's really quite simple," the Indian said good-naturedly. "In the Punjab, where I lived, there was a large tree behind my house. Each morning I would rise up early, tie my belt around it, and try to throw it down."

"A tree?" the boy marveled.

"For twenty years."

"And you did it?"

"No, little one," Gama smiled, "but after a tree . . . a man is easy."

The rising sun found the boy, as usual, with his buckets beside the wagon. Volanko tossed in his handful of sand. Almost imperceptibly the buckets had become a third full.

The boy practiced with diligence approaching fanaticism, training until he could not raise his hands even after putting the weight down. At night he sat cross-legged on his cot in the corner of the wagon, fascinated by an illustration in a dog-eared book.

"What are you looking at Yosselle?" the wrestler inquired peering over his shoulder.

It was a vivid depiction of a Biblical figure of colossal stature and shoulder-length hair breaking his chains and bringing down a pagan temple.

"Samson," the boy said. "He must have been the strongest man in the world."

"Continue as you have," Volanko said, "and someday you will be as great as Samson."

"*I* will?"

"What's to stop you? Place no limits upon yourself and you will have none." Volanko closed the book. "Train hard, but remember, Yosselle, to be a man is not about muscles," he said, affectionately putting his paw of a hand on the boy's shoulder. "These things impress you now. But in the scheme of things, it's hardly the beginning. Strength of character you don't get with a pair of buckets."

That night the door burst open, and over Yosselle's protests, Volanko pushed him out of the wagon and threw a blanket after him.

"No, no, you can't sleep here anymore. From now on, you sleep outside in the clean air. The climate is good here. Healthy for you."

The boy bedded down under the stars. After a while he rapped on the side of the wagon. "What if it rains?"

"It never rains here. . . ."

"Wild animals will come in the night."

"You like animals. . . ."

"Bugs will eat me."

"I killed them all last year. . . ."

"Thieves . . ."

"You have nothing to steal. Go to sleep."

After a long quiet moment, the boy slammed his fist on the wagon again.

"Volanko, why is it healthy only for me to sleep out?"

After a long silence the wagon door flew open and Volanko stormed out, tossing his blanket to the ground beside Yosselle. Mumbling Russian profanities he wrapped himself in the blanket and pulled it over his shoulders.

It was in Poona that Yosselle first observed yoga and made note of its postures. Volanko himself was in accord with many of its practices, especially fasting which he did one day a week. The flat fields near Poona were the ideal place for the boy to train. The climate was mild and balmy from October to February. By now, Yosselle was an able wrestler, and Poona gave him his first taste of actual combat. The local boys were no less serious than himself. When an opponent hurled him to the ground with such force that the dirt flew into his nose and mouth, he spat and came back, rising and falling, until he had swept his opponent off his feet and pinned him.

Volanko's words had become part of his consciousness. "Think that you're strong . . . and you are." Yosselle's image of himself began to change. Now he had the heart to rise when winded, to ignore cuts and abrasions, to refuse to quit even when doubled up in pain. He had learned how to endure.

After four months in India, the circus readied for its return to Europe. Yosselle perceived a change in Volanko; the wrestler became pitiless in his demands, pushing him beyond endurance.

The big man observed his exertions as the buckets were raised again and again, and still he prodded Yosselle on. "One more . . . one more . . ." And when the boy was past exhaustion, Volanko said, "One more . . . you can . . . you will . . . just one more."

Volanko imparted all the boy could absorb, building his pupil's wind, hardening his endurance and mental tenacity, polishing his Graeco-Roman technique. The boy detected a note of urgency in his master's voice, and sensed that time was growing short.

It was their last morning in Poona. The boy stood at sunrise beside the wagon. He was strong now, performing his morning regimen without strain or fatigue. The buckets now each weighed twenty-five pounds. Volanko added his fistful of sand. The buckets were full.

- - -

Somewhere in Eastern Europe in the blinding winter which began the year 1909, Volanko awakened in the early morning. Clearing the sleep from his eyes, he rubbed a peephole on the frosted isinglass window and peered out. In the dim light before day, he saw a figure standing bare-chested in the snow. His deep breaths became vapors which clouded the air. He rubbed handfuls of snow on his face and chest, accompanying the practice with invigorated oohs and aahs. The figure was Yosselle's. They had come full circle.

It had been a year and a half. The Issakoff Brothers' Circus returned down the road past Suvalk. And there, with affection but without fanfare, Champion Volanko deposited him on the road where they had first met. The boy stood before his mentor. Volanko looked down at the lithe, muscled athlete, a calm and different person from the frail youth who had been beaten half to death in the circus tent.

"Here we are, where we began," Volanko said, his steel-blue eyes masking his emotion. "I've done all I can, Yosselle. You're on your own."

The boy lowered his head sadly. As the wagon lurched forward, Yosselle ran beside it. "Volanko, take me with you."

"It's no life for you, Yosselle."

"*Volanko, I want to be like you!*" he implored.

"Good luck!" The wrestler waved his wistful good-bye as the wagon receded into the distance.

"Yosselle," he called back, "never forget who you are."

He would not see Champion Volanko again. But the lessons he'd learned would remain vivid for the rest of his life.

It was an April evening in 1909 when Yosselle Greenstein returned home from his year-and-a-half absence. His appearance had changed completely. He walked through town and several people he knew nodded without recognizing him.

He made his way up Malorotczik Street with mixed feelings of elation and concern, wondering if his family were still alive. There was a light on in the window. He entered the house. His mother, who was seated in a chair, looked up at him and fainted. His

brothers and sisters who were outside or in another room rushed in, wildly threw their arms around him, and began chattering all at once. His older brother Mottel cried.

Chaya Greenstein had mourned for her son from the time he had left. Thinking him dead, drowned in the nearby river where a whirlpool had taken two children the year before, she had recited kaddish, the mourner's prayer, and sat shiva, the traditional ten days of grieving.

For the next hours he sat with his mother, his brothers, and sisters, telling them of his travels and his time with Volanko. His mother alternately laughed, cried, or reached out to touch him. "Yosselle, Yosselle . . ." She said his name, hardly believing that he was alive and had come home. "Yet it's not you, little Yosselle. It's somebody who looks something like you."

When the emotions of his reunion had subsided, Yosselle turned his attention to someone he had often thought of during his days with the circus. "What would Leah Kaspersky think of me now?" he wondered.

Nine years before, at the age of seven, Yosselle Greenstein had set up "a business," an improvised swing with which to entertain the other children. Because none of his customers had any money, he took his pay in what seemed to him the next best commodity—buttons.

A five-year-old neighborhood girl had come with the other children. Like Yosselle, Rachael Leah Kaspersky was small for her age, with a sweet round face, soft features, and black hair. "Yosselle," one of the boys teased, "she looks like your sister."

She seemed shy by nature, yet she held her head high and looked at the others directly when listening or speaking. When she spoke, he liked the sound of her voice. And unlike the other girls who only giggled, when Rachael Leah found something amusing, she laughed out loud. She was small and delicate, and this brought out protective feelings in him. When she turned her dark eyes his way, a strange sensation overcame him. He stole glances at her as she ran or sat with the others. Rachael Leah wanted her turn at the swing, but had nothing to trade. So she went home and got some buttons. He swung her for hours, too shy to speak to her.

Her father was a short, deceptively plump but well-muscled expressman who made his living by hauling goods or grain to and from the train station with a horse and wagon. Arriving home for the Sabbath after his week's work and customary trip to the local bath house, he went to the closet to put on his best clothes, but found all the buttons cut off them. This usually cheerful man stalked out of the house and found Leah and Yosselle, along with his missing buttons.

From that time on, little Yosselle Greenstein was not terribly popular in the Kaspersky home, having acquired a reputation as a bad influence.

Yosselle would play near Leah's house in the hope of catching a glimpse of her. One day while he was enjoying pitching rocks with some other boys, he had just wound up to throw when Leah came out of her house all bundled up, her black hair blowing in the cold wind. The sight of her so unnerved him that his rock sailed through her mother's window. Mrs. Kaspersky came roaring out of the house, seized him by the ear, and dragged him home, whereupon his mother paddled him. The con man had added vandalism to his reputation.

Despite such mishaps, by the time she was eleven and he thirteen, each of them had overcome their shyness. They walked together in the fields around Suvalk, where Yosselle picked wild flowers for her. In appearance and temperament, one soon became a counterpart of the other, their affection so precocious that it seemed almost adult.

Now Yosselle Greenstein had returned to Suvalk as a young man of sixteen; he walked through the town past the square and saw her at the "sashaygas," the promenade lining the park where they used to hear the military band play. Leah too had shed most of her childhood.

They waved to each other from across the promenade. He went to her.

"Leah, you recognized me?"

"Of course," she said. Half an hour later they decided to get married. Their parents would hear none of it. They would have to wait. Meanwhile, Yosselle took a job in the tannery where his father had worked and began to readjust to life in Suvalk.

His family, friends, neighbors lived as they always had, but his time with Volanko had changed him. As Volanko had predicted, he had acquired more than muscles. He looked at the lives of those around him. "What future is there in this place?" he thought.

Suvalk was usually peaceful and a better place than most in Eastern Europe; yet even here rumors of impending violence against the Jews, which was epidemic elsewhere, made life uncertain.

One day on his way home from the tannery, Yosselle watched as one of his neighbors, an old man, boarded up his house.

"Simon . . ."

"Yosselle, run home. Tell your mother . . ." His voice was filled with quiet terror.

"What's wrong?"

"Pogrom!" the man whispered, and disappeared behind his weathered door.

Some days later, while laboring at his bench in the tannery, Yosselle heard a peasant workman talking loudly about the pogrom, the Jew massacre that was, sooner or later, bound to come to Suvalk.

"I'll do my part when the time comes," the man bragged. "We'll sweep the kikes out with an iron broom."

"This man would willingly murder me, my mother, brothers, sisters, and Leah. But what have we ever done to him? What?" Yosselle wondered.

Yosselle had no ill feelings toward Poland or the Polish people, and had friendly dealings with many. He was not sophisticated and knew little of politics, but he had returned from India with a new life and new attitudes. He could no longer accept the life of a victim, which would be his lot in Eastern Europe. In the synagogue he had learned the basic principles of Jewish life. "And you shall love your neighbor as yourself." "Thou shalt not murder." "But," Volanko had said, adding a teaching of Rabbi Akiva of old, "if your neighbor shall come to slay you, rise up early . . . and slay him first."

In April of 1910, just before Passover, and in the flurry of activity preparing for the holiday, Yosselle found himself up on a ladder whitewashing Leah's house, but turned as he heard a commotion. A soldier lurched down Zagorodny Street, running erratically with a drawn sword in his hand, its blade dripping blood. A crowd pursued him: "Stop him! He just killed somebody!"

The crowd ran after him with caution, as every so often he turned and lashed out with his weapon to keep them back.

In the village, for no apparent reason save too much vodka and a convenient victim, the soldier had struck an old Jew with his blade, taking the top of the man's head off and killing him instantly.

The young wrestler came down off his ladder and, skirting the back of the house, trailed the man to the lakeshore, a place that he knew well. The soldier, who was in his late twenties, was not very tall, but strong, with thick limbs and a broad peasant face. He stopped long enough to puke up his vodka, then looked around, and having evaded the crowd attempted to sheathe his blade. At that moment, Yosselle charged out of the bushes, struck him with both hands across the back, and tackled him. The soldier went down with his blade halfway in its scabbard. Yosselle stepped on the weapon, broke it, and came up with the sword's heavy brass hilt in his hand. As the man rose and whirled to stab with the sharp remnant of the blade, Yosselle whipped the hilt around, and the improvised brass knuckles slammed into the side of the soldier's head. He danced on rubber legs for two steps, then his eyes rolled back white, and he pitched forward in a heap.

Yosselle stood over him, breathing hard, waiting for him to get up; but he lay still. The young wrestler knelt and felt for a pulse. There was none. He bit his lip and again he felt at the heart, the neck, the wrist . . . nothing. A fly danced on the soldier's eyelid, but it did not flutter. Yosselle looked around. No one had seen. He darted quickly away, waded through waist-high water for a couple of hundred yards, and circled around to the far side of town. No one would see him coming from that place.

His sister Chani, her head wrapped in a babushka, was cleaning the house as he entered from the back street, his clothes soaked and soiled.

"Yosselle, look at you. What have you been doing?"

"Chani, I . . . just killed a man."

Her hand flew to her mouth. "Who did you—? . . ."

"A soldier." He told her the details.

"Yosselle, run away now"—she began to panic—"run!"

"Chani," he said, "if I leave, they'll know it was me. Clean the house, say nothing."

He remained in town, but had to remove himself from sight for a while. The only hiding place he could think of was the chicken coop behind the house. He hid himself there under gunny sacks, not making a sound, taking meals, or even moving, though lice swarmed over him. After two interminable days and nights he emerged and went to the bath house, where he had his clothes boiled and himself deloused; then he went back to the house for Passover and waited.

As Yosselle Greenstein took part in the retelling of the liberation of his people from bondage and their preservation from the slaughter of the first-born in Egypt, army officers were in town asking questions about the soldier's untimely death. If they discovered it was he who had done the deed, doubtless there would be no talk of circumstance or justification; it would be the firing squad or worse. Would they harm his family? Would it create an excuse for a bloody riot? He did not know, but sat at the Passover table hiding his desperation and reading the traditional words aloud.

Each evening he sat at the table in a cold sweat, waiting for the knock on the door that never came. When the days of Passover had ended, Yosselle went about his business with renewed enthusiasm. In Suvalk, Poland, in April of 1910, the Angel of Death had passed over.

At this time Yosselle had two preoccupations in life—Leah and wrestling. He continued daily training, lined up frequent matches behind his house, and won all of them. He then challenged the best Suvalk had to offer, Issac the blacksmith, a bushy-browed man seven years older, forty pounds heavier, and half a foot taller. Surprising even himself, Yosselle pinned him with his style of quick attack, polished technique, and never-say-die tenacity. After that match he acquired the nickname "malchick" which loosely translates to "kid" . . . "Kid" Greenstein.

As he approached his eighteenth birthday, Yosselle asked his

At a penny arcade in Suvalk, Joe at the wheel

brother Mottel to act in place of their deceased father and arrange things with Mr. Kaspersky. Over the years the attitude of Leah's father toward the little button thief and vandal had mellowed to something akin to paternal affection. To Mr. Kaspersky, Yosselle Greenstein was obviously neither a prince nor a scholar, but he was an honest and hard-working young man, and totally devoted to Rachael Leah. For her part, she would not even look at anyone else. Mr. Kaspersky consented to the marriage.

At eighteen and sixteen, they were married simply, presented with Leah's small dowry, a pair of candlesticks and a tablecloth. In the early morning after their wedding night, Yosselle's child bride awakened to find her groom staring out the window.

"Leah, my mother sometimes gets a letter from her sister and her husband in America. . . ."

"You want us to go to America." Her intuition cut him short.

"Yes."

"How can we leave? With what?"

"I'll go first and send for you. My aunt and uncle live in Texas."

"Where?"

"Houston, Texas."

"Who else do you know there?"

"I'll meet people."

"You can't speak English."

"I'll learn."

"How will you live?"

"I'll work hard, and save. . . ."

"Yosselle, I don't know. . . ."

"What's to know, Leah? People say that a man can be a man there, and not fear his neighbors."

LEFT TO RIGHT: Leah, Chani, Yosselle, one week before he leaves for America

"Others say that families live in the slums of New York ten to a room, with disease and rats. Is that better than here?"

"Yes. I think a man can change his life in America."

Finally she agreed.

They had been married only eight weeks when Yosselle packed his bag, and he and Leah went to his mother's house. He embraced her, Mottel, Chani, Sara, Mikhail . . . and though they were in tears, they did not try to dissuade him. Mottel kissed his little brother and said everything that could be said: "Be well. God watch over you."

He left the ramshackle house on Malorotczik Street, and it seemed to him the saddest moment of his life. He feared that he would never see his family again. He and Leah walked to the train station. "You'll send for me . . . ," she said with a veneer of cheerfulness. Out of concern it would prevent his going, she did not mention that she was pregnant. "How long, Yosselle?"

"Soon."

He kissed her good-bye and held her, then offered his encouraging boyish grin for her parting view of him as he swung onto the train, leaving the squalor of Suvalk behind for the last time. "Soon, Leah, soon." He waved and the train steamed toward Germany.

At the Bremen docks he booked passage on the *Frankfurt*. The voyage across the North Atlantic was a stormy one, and in his steerage berth beneath the ship's waterline, he lay deathly ill for the first three days. Gaining his sea legs, he volunteered for work in the galley for something to do. He peeled potatoes for the ship's cook and was rewarded with choice meals, which, with the ocean air, quickly replenished his strength. After three weeks at sea, he stood at the railing with hundreds of other immigrants as the ship glided up the Delaware River to unload cargo and some passengers at Philadelphia. At the second port of call, Yosselle Greenstein stepped off the *Frankfurt* onto the free soil of Galveston, Texas, U.S.A. It was June, 1911.

A BULLET FOR THE JUNKMAN

His first view of America was of a charming and languid port city; he walked wide white boulevards of oyster shell which were lined with the red and pink flowers of oleander evergreen. Palm trees swayed in a tropical breeze from the Gulf of Mexico. Galveston was stately mansions and manicured lawns, oil wells burning in the far distance, dirt streets and clapboard houses, magnolia and heavy humid air, black laborers shiny with sweat hauling cotton bales on the docks, horses vying with motorcars.

Yosselle Americanized his name to "Joseph," and with a bundle under his arm and a piece of cardboard marked "Silverstein, Hutchinson Street and Harrisburg Road, Houston, Texas, U.S.A." tucked into his hat, he took the Santa Fe Railroad to nearby Houston.

He was directed to a neat but undistinguished house. Though he knew it to be the home of a working man, it was a palace compared to the dirt-floor hovels of Suvalk. He knocked on the door of the porch where his Uncle David, a benevolent hawk-faced man who coughed when he laughed too hard, was snoozing. Through the open door, he could see his Aunt Rachel, his mother's sister, laboring in the kitchen. Spying her nephew, Rachel paused just long enough to awaken her husband and rushed to clutch young Joe to her bosom. Uncle David and Aunt Rachel sat him down for a meal, and brought out a bottle of homemade "schnapps" to toast the occasion. Joe declined a second toast because the first, imbibed in the Texas heat, had already gone to his head. After a few hours' reminiscence, Uncle David presented him with clothes which, though ill-fitting, were more suited to the climate. Joe was bedded down in a spare room. The next morning after eggs, potatoes, homemade bread, and fruit, a breakfast Joe imagined would put the Czar's to shame, he accompanied his uncle to work.

David Silverstein's livelihood, a delivery route for the Prince Bakery, was made immeasurably easier by a horse who knew the route so well that Uncle David never had to start and stop him. This remarkable animal just went from house to house where the deliveries were to be made. Joe was impressed. "What kind of horse is that?"

Uncle David, who knew as much about horses as he did about brain surgery, replied with unabashed pride, "That, Yosselle, is an American horse," before countering with his own inquiry.

"So, your plans?"

"I want what the horse has," answered Joe. "Work."

"A pity you don't have his qualifications," said Uncle David. "The horse understands English."

The next day, at Uncle David's suggestion, Joe took to the easiest vocation of the new immigrant—peddling. With two baskets of apples which he had scrupulously polished, and an English vocabulary grown to three words—"Apples!" and "yes, ma'am"—he set out.

He walked the streets of Houston and, after a while, saw a woman sweeping her front steps.

"Apples!"

She looked up and said something. Not knowing what the words meant, he took it as a sign of possible purchase. He was so elated he hadn't noticed his muddy shoes. He walked up onto her clean steps, grinning from ear to ear. The little woman took one look at his dirty footprints and swatted him in the head with her broom, knocking him and his two baskets of apples into the street. After a few weeks of trying, Joe accepted his first venture as a failure.

Fortunately, in the streets of Houston, Joe had met another immigrant, a tall older man with an aquiline nose and thinning hair named Benny Biskin. Biskin, an itinerant peddler of dry goods and notions, was making a living by selling everything from work clothes and blankets to needles and thread. His customers were farmers who valued him because he brought them quality articles at fair prices, would take special orders and even extend credit. However, Biskin's business required that the three hundred pounds of goods he transported be loaded and unloaded with some frequency. For three dollars a week, Joe agreed to take on the heavy work.

Down dusty roads with the horse and wagon, across the rich cotton fields of the Blacklands and seemingly endless plains, they ambled past barbed wire and whitewashed farms, ringing a bell to announce their arrival. Joe found his traveling companion to be a quiet man, whose character had been molded in the crucible of Eastern Europe, though Biskin didn't mention exactly where he was from. He never had a meal without a prayer before and after, and he ate without speaking, as if at any moment he expected it to be taken away. With a philosophical mien, Biskin would sit silent for a time before nodding to himself and musing out loud, "Yes, America" to no one in particular. Joe knew the man had infused the whole of his bittersweet life into those two words: "Yes, America."

When Joe spoke to Benny of his enthusiasm for wrestling, Biskin took it upon himself to act as his "promoter." Wrestling was as popular in Texas as it was in India, and although the Texan variety was less traditional, it was pursued with the same enthusiasm. The rules of Texas "wrasslin'" were simple: be sportsmanlike, shake hands before and after the match, and try not to leave your opponent deaf, blind, or sterile.

On Sundays, after the farmers had attended church and Benny had concluded his business, he would casually say, "Mr. Jones, Joseph here likes to wrestle a little. Do you know of any boys around here who are like-minded?" It was a superfluous question in these parts. So Joe "wrassled" the strapping farm boys while onlookers rooted for their relatives, and Biskin called out pointers in between sips of lemonade.

The days of traveling with Benny Biskin through the Texas back country were broadening and pleasant and would have been more so had Leah not been constantly on his mind. His worry over her welfare did not show in his letters: "Dearest Leah, my loving wife. All is well. In this envelope you will find eight American dollars. . . ." At his munificent salary of three dollars a week, it would take forever to bring her over. But how could he ask a poor man like Benny for a raise, when he felt guilty accepting three dollars' pay?

They were on their way back to Houston after a week on the road, and Benny seemed unusually pensive. "Joe, what do you do with your money?" he asked.

"I spend a dollar a week, and save two for my wife."

"That's what I thought," Benny said and did not say much more all the way back.

When he stopped the wagon at the Silverstein house, Biskin cleared his throat as he handed Joe his pay.

"Thanks Benny," the young man said.

"Joe, you're fired."

"What did I do wrong?" Joe asked with injured feelings.

"Nothing. I feel toward you like a son"—Biskin opened his hands in frustration—"but I'm a peddler of shirts and underwear, a seller of blankets and suspenders. My living is a penny here, a penny there, a nickel, a dime. For me, it's enough." He pushed his battered hat back on his head as he struggled for the words. "I don't want to hold you back. You need something better. This is America, and you'll find your way. So, you're fired"—he leaned down and shook his helper's hand—"*and good luck.*"

"So long, Benny," Joe said as the peddler continued down Harrisburg Road with his horse and wagon and bell clanging.

After a stint as an assistant tinsmith with the Houston firm of Blumenthal Brothers, Joe took a job at the Wesley Blakely Ranch near Alvin, Texas, to try that most American and romantic of occupations, cowboy; but instead of adventure, he found tough, dirty labor for room and board and thirty dollars a month. After two months of drudgery, of cleaning stalls and digging postholes, he "moseyed down the trail" and returned to Galveston to try his luck there.

More and more immigrants were coming to Galveston, in preference to the crowded ports of the East, and many were Jews. Rabbi Henry Cohen, the leader of the city's Jewish community, took it upon himself to find homes and jobs for the newcomers. Concerned more with results than social convention, Rabbi Cohen sped about town on a bicycle.

In Galveston, hunting for a job opportunity, Joe was advised to see the rabbi. The young immigrant stood on a street corner when he saw the wiry man in a black suit racing down the street toward him on a bicycle.

"Rabbi!" Joe shouted as he stepped off the board sidewalk and flagged him down. The bicycle careened around him broadsliding to a stop in a small cloud of dust.

"Rabbi, I can't speak English," Joe explained, "but I need work."

"Have you tried the docks?"

"No."

"Maybe there's something I can do."

"My dearest Leah," Joe wrote in his next letter. "I am now in Galveston working on the docks of the Southern-Pacific Railroad where the tracks end at the Gulf of Mexico. I have seen some of Texas and have met many kind people. I know I have done right coming to America. In this envelope you will find fifteen American dollars. . . ." He earned this money with a longshoreman's hook slung over his shoulder, laboring for thirty cents an hour . . . three dollars a day.

His evenings, like all the others since he had arrived in America, were spent quietly, laboriously teaching himself English. He had quickly learned the alphabet, but it seemed to him that the alien words were rarely spelled as they sounded. He kept a children's pulp picture magazine in his back pocket for its visual reminders of what the words meant, and progressed to reading the copy in newspaper photo ads for food and clothing. Then, late into each night with a dictionary beside him, he attempted to absorb whole pages of discarded copies of the Houston *Post,* as if that tenacity that Volanko had infused in him would also serve to master stubborn letters. Though his penmanship was little more than chicken scratch, and his attempts at phonetic spelling were laughable, Joe found that his ear for music also applied to language, and he memorized words and phrases by constant repetition until he had internalized them.

Sometimes there were moments of amusement. The dockworkers staged "races." Two 500-pound bales of cotton were loaded onto a pair of handcarts. From a standing start two men, each pushing a cart, would try to beat the other in a foot race. Joe, who excelled at this grueling game, became the local champion. At odd times on the wharf he would practice wrestling exercises, clutching a heavy sack of beans to his chest in a bear hug, squeezing with ever-increasing tension, conditioning his arms to become a vise.

His Sundays were spent at Galveston's Murdock Beach where amateur wrestlers competed. It wasn't long before Kid Greenstein was whipping all comers in these free-style "catch-as-catch-can" events. His speed, polished attack, and tenacity proved too much for men of greater size but less skill. During the week at lunch and dinner breaks he wrestled dockworkers on the grass near the wharf and was never beaten. He began to build a reputation.

One early afternoon after five hours of loading grain, for his workout Joe engaged a larger man in a match, and, after pinning him, sat down to have his vegetarian lunch. He looked up to see the dockworkers excitedly gathering around a large black man who had sauntered onto the wharf: in his early thirties, six foot one, about 210 pounds, dark-skinned and with a broad nose, angular face, high cheekbones, and a strong jaw, the Negro's cleanly shaven head resembled a cannonball. His shoulders, heavy as beef sides, threatened to burst his elegantly tailored clothes. He might have looked sinister when he smiled with his uneven teeth, but had instead a surprisingly pleasant and gentle demeanor.

"Hey, Kid!" Curly, the boss of the dockworkers, called for him to come over, and he made his way through the small crowd to where Curly was standing with the black.

"Kid, I want you to meet Jack Johnson." Johnson had been Heavyweight Champion of the World for three years, and on his way through Galveston returned to the docks of the Mallory Line where he had worked.

"Whaddya say, Joe?" Curly touted one against the other: "Think you can beat him?" Johnson too had been a champion racer of the Galveston docks.

"Well, it's been a while . . . ," Johnson said in a cultivated voice as he stripped off his jacket and rolled up his sleeves. A pair of 500-pound cotton bales were tripped onto their carts. Joe reluctantly took hold of the oak handles. The contest would be fair as, despite the pugilist's weight advantage, he was out of practice, had not warmed up, and his smooth patent-leather shoes would give him little traction.

"Go!" Curly shouted and Joe, straining to get momentum, could hear Johnson's feet slipping. Step by step they moved faster, and though Joe had gained the advantage, the Champion's long legs quickly covered lost ground. Running with the carts while straining to support and steer the load, they struggled across the finish line a draw.

Johnson regarded him with amused curiosity. "Where you from, Kid?" he asked.

"Poland."

"Have you ever heard of a skinny Jewish fellow by the name of Choynski?"

"No."

"He's retired," Johnson went on, "but he was a fighter . . . always fifty pounds too light, but he fought all the greatest heavyweight champs . . . Sullivan, Corbett, Jeffries, Fitzsimmons." Joe's blank expression indicated that he had never heard of any of these men.

"Most of them said that Joe Choynski was the hardest hitter they ever went against. You know what Choynski did to Jeffries back in 'ninety-seven? Hit him so hard in his mouth that Jeff's lips were plastered back between his teeth and *had to be cut free with a knife.* That's a fact." Johnson wiped away a few drops of sweat from his massive head. "Why even Corbett said that he took more punishment from Choynski than he got in all his fights put together." The young man's puzzled expression never wavered. "You're wondering why I'm telling you this? I'll get to that.

"Choynski and I had a bout right here in Galveston in 1901, just after the flood. I was strong and had been fighting a couple of years, so I didn't pay his reputation no mind, because like I say, he didn't weigh but one sixty-five and he was old, past his best at thirty-three. You know what happened?"

"What?"

"He knocked me out!" Johnson laughed unashamedly. "Well, it was only three rounds but fighting was illegal in those days, so the bulls jailed us. While we were in the lockup we boxed, talked boxing, and boxed some more. So you see, I knew him real well." He paused.

"Now, Kid"—he eyed Greenstein squarely—"when I came up on the wharf I saw you pin that big man to the ground, and you just damn near beat me at my old game. You look like Choynski, walk like him, have the same style, and the same first name. So stick with it because what Joe Choynski had . . . *you have that in you.* I know what I'm talking about, Kid," he said with finality. I'm the Heavyweight Champion of the World."

"Mr. Johnson . . ."

"Jack . . . you call me Jack."

"When my wife comes from Europe"—Joe offered his hand and his best English—"you come to dinner in my house?"

"Why, I'd be pleased Kid. I'll make a point of it."

Curly, taskmaster of the Southern-Pacific docks, impressed by Joe's wrestling, his showing in the cart race against Jack Johnson, his level of conditioning, and his long arms, was Kid Greenstein's most enthusiastic advocate. At first glance Joe's arms appeared normal, but actually they were abnormally long; Joe could almost touch his knees while standing erect. Though he was 5 feet 4½ inches tall, his outstretched arm span from fingertip to fingertip reached 6 feet 1 inch.

"With your power and a reach like that," Curly wheedled, "you'd make a hell of a boxer."

"I know nothing from boxing," Joe explained, "I don't think I'd like it."

"How do you know 'till you've tried? You're throwin' away an opportunity here," Curly insisted.

Johnson's words of encouragement, combined with Curly's talk of money—"I'm tellin' ya, Kid, there's bucks in it"—made Joe begin to wonder if maybe there wasn't something to this boxing business. He considered it. "Okay."

He let himself be promoted into a boxing match at the Beavers Club at Twenty-fifth and Market the following Saturday night. He found himself in the alleyway outside the club, dressed in an oversized pair of trunks, with Curly lacing on his gloves. Although Curly was full of assurances—"It's all set, Kid. Don't worry. Just use your head. You're a natural"—he was more concerned with his share of the purse than with any details of the art of pugilism.

Kid Greenstein entered the club, where a primitive rope ring was surrounded by straw hats; cigar smoke hung over the place like a gray halo. All eyes turned to him, then the buzz of voices grew louder as the wagering began. Joe scrutinized his opponent from across the ring. The Irish boy, about nineteen, wore his ginger hair slicked straight back, and with his close-set eyes and a plethora of freckles had an undistinguished face, except for his nose which was flat and his ears which resembled small cauliflower bouquets—unusual traits for one so young. Otherwise, down to his skinny hairless legs, he appeared no different from other sons of Irish workmen whom Joe passed daily on the streets of Galveston.

A referee in gartered shirt sleeves called them out to the center ring and announced their names, they touched gloves, and a bell was rung with a wooden mallet. The Irish youth advanced and Joe decided to feel the situation out . . . whack! A white light went on over Joe's left eye as the Irish boy split his eyebrow open. The crowd roared its approval, but Kid Greenstein was not really aware of what had happened. He watched, almost as a spectator, as his opponent shuffled in and delivered two jabs and a quick right, the combination bouncing off Joe's head as if it were a speedbag. Joe had not yet thrown a punch. He had no idea how to defend himself without grappling; and, though hampered by the gloves, he tried this in the clinches. The boxer smacked him in the kidneys with his bony fists. "Booooo!" Catcalls flew. "Knock the bum out!"

The crowd, uniformly disgusted, smelled an untimely end to Greenstein. It dawned on Joe, "I am strong. I am quick. *I am a wrestler. What am I doing here?"* The little Irishman did a spritely two-step and with a feint and practiced hook—whack!—split his other eyebrow. Joe began to feel warm blood running down his face. So much for compliments; so much for tales of Joe Choynski; so much for confidence without preparation. In the corner, Curly looked dumbly at him and winced as the little Irishman's leather pistons pumped furiously to the head and body. "Curly, you bastard . . ." The crowd began to throw things. "I have to get out of here . . . have to do something. . . ." A bell went off in his head as the boxer planted a right hook on his ear. Wrong bell. "There's a way . . . think . . . think . . ."

Joe weaved back, holding his elbows high. "There's a flaw . . . look for the flaw. . . ." He saw it. When his opponent threw the hook, he momentarily left his chest unguarded. Joe covered his head and moved away as if ready for the end. The boxer waltzed in and threw the hook; Joe waited for it, let it glance off his elbow, then—with everything he had—stepped forward with a straight right hand to the solar plexus. Bull's eye. At the moment of impact, Joe thought he saw the Irishman's eyes pop out from the blast of air that backfired into his lungs. The freckled youth dropped his guard and heaved to catch his breath. Joe half closed his eyes and desperately swung his whole body into his right hand. The boxer saw it coming and threw up his hands to block, but too late. Kid Greenstein's arm sailed through his defense, the whipping roundhouse catching him squarely on the nose and taking him off his feet. The little pugilist landed on his seat still heaving for air, but he was out.

In the corner Curly came alive, screaming congratulations. Joe stood over his fallen opponent, as blood dripped from the Irishman's nose. He had neglected to tell Curly that he couldn't stand the sight of blood. He vaguely heard the referee counting, then wavered a moment before passing out. The match was declared a draw.

Later, revived, but with a terrible headache, he got dressed in the alleyway. Curly handed him nine dollars along with a frank admission. "Kid, you stink; you'll never be a boxer."

It was Joe's last professional boxing match. He returned to wrestling.

■ ■ ■

On Sunday, a week later, Joe relaxed on Galveston beach looking out over the blue Gulf tidewater. Hours before, he had pinned three wrestling opponents in quick succession.

He sat with a steel spike and from time to time would flex his hands against it, an exercise for the mind more than a test of strength, as his hands were not equal to bending it. Sometimes he would hear a voice from within say, "You can . . . you will . . ." and the feeling came over him, an inspiration born of Volanko's promise that there was nothing beyond doing if only one had the courage and the patience. In these rare moments, when he flexed his hands, the spike gave ever so slightly.

Joe was ready. He decided to turn professional. There had been other motivation to do so; Leah's last letter had informed him of her pregnancy.

Kid Greenstein's first professional wrestling bout took place in the oil fields near West Columbia, Texas, about fifty miles southwest of Galveston. Before an audience of roughnecks and roustabouts who sat atop wagons or nearby drilling rigs, he was to wrestle a veteran named Frenchie, best two out of three falls for a purse of twenty dollars—winner take all.

As an added incentive, before the match, Joe had written a letter enclosing the purse money to Leah and stamped the envelope.

Frenchie was a bulldog of a man with a slight paunch, one gold tooth, and a penchant for choking, head butting, and eye gouging. He offered Joe a terse greeting: "I'm gonna tear off yer head, and throw it in yer face!" Joe pinned him for the best two out of three falls in an hour.

After that, Kid Greenstein was booked at Houston's Moose Club against Jake Rudnick, a wily and experienced matsman, five years older and twenty pounds heavier. He beat Rudnick by applying the bear hug. After the match, Rudnick looked at this young man so much smaller than himself.

"Kid, you nearly squeezed the soul out of me."

"You know how I do it?" Joe volunteered.

Rudnick waited with interest.

"Beans," Joe said.

Kid Greenstein wrestled in quaintly named places such as Goose

Creek and Burkburnett, in the oil fields and dance halls of Texas and nearby Louisiana. He wrestled for all types of promoters, some gentlemen, some shady. He learned soon enough that no one was giving anything away; one of his most outrageous surprise opponents was a man with fingers the size of a normal man's wrist and dubbed the "Buffalo Butcher Boy." Fortunately, technique overcame tonnage. For such combat the purses amounted to whatever was offered, usually ten to twenty-five dollars.

These wrestling bouts were sporadic and he was not yet a union stevedore. Men in his position could be hired only after all the union men had been put on. He would get up at four A.M., and after breathing exercises and calisthenics, would be the first one at the dock; but still, he could get only one or two days' work a week. If the boats were late, and didn't come in that day, hired or not . . . he was out of luck. While the other men would retire to the local pool hall, he could not. Leah was waiting.

A friend had told him that a job as a deckhand on one of the tramp steamers was a more reliable position. It paid a dollar a day every day, with free food and lodging. A boat was shipping out the next day bound for Japan, and they needed men.

Joe was approaching his nineteenth birthday when he shipped out for Japan. He arrived in Yokohama in the spring of 1912. It was the last year of the Meiji Period and the rule of Emperor Meiji Tenno, a time of increasing Western modernization. Japan was a land where tradition survived in the midst of change. It had been scarcely thirty-five years since the feudal Samurai knights had been abolished as a class, and an edict issued which forbade the wearing of their swords. Yet so advanced had the military might of the Rising Sun become that only seven years before Joe's arrival, they had sent the Russian navy to the bottom of the Tsushima Straits and slaughtered her Manchurian forces during the Russo-Japanese War.

In Yokohama of 1912 the vestiges of the *bugei,* the samurai martial arts, remained, though they had been tempered from use in war to less bloody martial ways of personal discipline, and would in the years to come evolve into competitive martial sports like Judo. During his days with the circus, Joe's curiosity had been aroused as he heard Volanko speak of the Japanese physical arts. In this land of small men traditional systems of self-defense had been created whereby through knowledge and training a man could overcome one or many larger opponents. It was the efficiency of the Japanese wrestling of which Volanko had spoken that most interested him.

Kid Greenstein found Yokohama fascinating in its contrasts. A seaport town known for its silk trading and prostitution, Yokohama was a cosmopolitan city which attracted foreigners: Portuguese, Dutch, Eastern European Jews, White Russians.

Upon arriving, Joe began stopping Europeans in the street, attempting to communicate. "Farshtaist Yiddish?" he inquired in his native tongue. Nothing. "Excuse me, can you speak English?" Nothing. He tried the same routine in Polish, Russian, and his smattering of a few other languages. Those he encountered knew none of these and backed away from him; when he resorted to pantomime to demonstrate his interest in wrestling, his contortions made him look like a maniac.

Finally he stopped a likely candidate, a bearded young Caucasian. "Farshtaist Yiddish?" he inquired again.

"If I can't understand Yiddish," the blackbeard quipped fluently, "then there's no sense in my mother talking to me, now is there? What can I do for you?"

Joe explained that he wished to be taken to a place were Japanese wrestling was practiced. After consideration, the young man led him through a maze of narrow lanes, past wood-and-paper houses and small shops to a back street where he pointed to a small single-story building and then departed. The dark brown wooden structure had the panels of all four walls patterned in a simple motif of squares and the corners of the roof turned up in the antique mode; and though the place was old and without ornament, it was well cared for, and the street out front was freshly swept.

Guardedly, he walked to one of the sliding wooden walls, a doorway which had been partially opened to admit air, and peeked in. It was a large, unfurnished room with thick rice-grass *tatami* mats butted one against the other and inset into the floor. In the center of the chamber was a squat Japanese man who stood in bare feet wearing a flowing *hakama,* a black training uniform, a starched blouse with loose sleeves and a stiff floor-length masculine skirt divided in half into pantaloons and tied at the waist with a black sash. The man in black was short and rather heavyset for a Japanese, but his hands and feet were small. His round face was almond in color, with extremely slanted eyes and completely hidden eyelids which imparted an almost Chinese appearance. His black hair was thinning, but his face did not have a line on it, and Joe could only estimate his age at fifty.

The place was soundless as the headmaster stood before ten students, all about twenty years of age, dressed in the traditional uniform, and arrayed in a single neat line. Like ten matched dolls they sat on their heels with legs tucked under and backs kept rod straight, holding their hands partially open while folded on their laps, with right and left thumbs and forefingers touching to form a circle. Their eyes were closed and they meditated for a time in this manner, facing the teacher and the small wooden shrine on the wall behind him. The polished wooden rectangle contained a small incense urn and lacquered wooden plaques of Japanese writing.

"Moku-so-yame," the master said concluding the meditation, and twenty pairs of eyes snapped open in unison. The students leapt to their feet and lined up in pairs facing one another. Aware all the while of the foreign face peeking in, the instructor strode across the mat and stuck his head out the doorway, staring impassively at Joe, nose to nose. Taken aback, and not wishing to give the wrong impression, Joe spoke slowly (as if that might help), and augmented his speech with hand gestures.

"I . . . would . . . like . . . to . . . watch. May . . . I?" He fully expected to have the door shut in his face, but to his surprise, the headmaster motioned for him to remove his shoes and enter. Once inside, he was signaled to sit down in the back of the school. With all this formality he wondered what kind of wrestling they did in this place. He wondered if he was in a temple and not a gym.

For the next several hours, he observed as the students practiced rolls and falls, blocking and striking with arms and legs, grabbing and evading and throwing one another. While it was all done efficiently and with dedication, and the movements appeared effective, it was a melange of seemingly disparate actions.

"How do they put all of this together?" He did not have long to wait for the answer.

The headmaster spoke and the students hurriedly lined up. In his school everyone bore the most serious attitude, and everyone ran, even jumping up or sitting down in double time.

Like an officer reviewing sweaty troops he eyed his charges sternly, then walked ten paces to the center of the mat, turned and bowed to them, and they to him.

"Hajime," he said, and suddenly his students rushed at him in twos and threes. But when they closed with him to grab or strike, he quickly upended them and using their own weight and momentum hurled them to the mat with a "whump" that echoed through the place. When a student seized his wrist, the headmaster evaded the grasp and the attacker's own wrist became the target; the torso which was attached to it was cartwheeled into the air or forced down by the pain of the pressured joint. Encircling, they came at him, and the only sound to be heard was bare feet padding hurriedly on tatami, climaxed by the thud of a redirected and fallen body. Each of the young men had been schooled in the art of falling, and softened the shock of landing with a slap of the arms on the floor at impact or a roll-out to avoid injury. Never once repeating himself, the teacher employed throws, strikes, kicks and footsweeps, grappling, joint locks, and *atemi* pressure-point attack, as he tossed them around by their arms, hands, fingers, ears, the scruff of their necks. Now the pieces all fit. The man was a master of his business, and master of anatomy as well, a spellbound Greenstein realized. He was a dancer whose stance was rooted in the very earth beneath his school, yet whose light steps evaded all confrontation. He could immobilize with a flick of the hand. An attacking student was lightly touched at a hidden pressure point on his abdomen, and he doubled up, whereupon the teacher kicked his legs out from under him, and dropping to one knee, spun him into the air, all in less than two seconds. His control was remarkable. He could be as gentle or as devastating as he wished, and this was only practice. If he meant his students harm . . . "God help them," Greenstein mused.

This Japanese magician seemed neither to exert much energy nor to concern himself with the violence around him. His face never varied from its expression of tranquillity as each deceptively simple movement was executed with a startling grace and economy of action, as if he possessed a keener sense of time. There was a religious quality to this art, Joe understood intuitively. The man had studied and trained his body, but it was the spirit that was at work. This was not wrestling, he concluded, and wondered what on earth it was.

It was *Jujutsu,* vulgarized to "Jiu-jitsu" in the West, an artifact of samurai combat which had been practiced within 725 systems dating back to the mid-sixteenth century. Jujutsu translates to "gentle art," but "gentle" is intended to mean "pliable" or "adaptable," as a strong oak sways in the wind.

"Yamé," the instructor said, and the students, none of whom was injured, ran quickly, straightening their uniforms, and arranged themselves in the neat line again. They took the seated posture, bowed formally to the shrine on the wall, then to their teacher, and the class was dismissed.

Joe was convinced that he was meant to learn this art. While understanding from Volanko that complete mastery would require half a lifetime, he knew that even brief study, when combined with his present skills, would pay handsome dividends on the way to a lightweight wrestling title. Leah was still waiting, but he felt that this was important and that she would understand.

The teacher stood at the door looking after the last of his students. Cautiously aware of Japanese suspicion of foreigners, Joe approached him, and spoke slowly and deferentially, using hand motions.

"Sir"—he bowed as he had seen the others do—"I . . . am . . . from . . . America. . . ." To his surprise, the teacher now seemed quite open and friendly, the authoritarian bearing with which he had taught the class having vanished. "I . . . am . . . a . . . wrestler . . . ," Joe continued somewhat encouraged, "Graeco-Roman . . . free-style . . . catch-as-catch-can. . . . A wrestler . . ." He attempted his ridiculous pantomime again, but this time it worked.

"Hai, wakarimashita." The Japanese nodded his understanding, amused by the display.

"Sir . . ." Joe pointed to the mat. "Please . . . I . . . want . . . to . . . learn. . . . Will . . . you . . . teach . . . me?" The almond face turned enigmatic again, the Japanese pulled on his ear thoughtfully, as if old and new were contesting within him. Discreetly, he sized Joe up to ascertain his fitness for such an undertaking.

"Teach . . . me," Joe said again, and the man in black nodded impassively, raising a finger to his nose, the custom to indicate oneself. "Yamashita Sensei," he introduced himself. Yamashita was his family name; Sensei his title—teacher, guide, Master of Jujutsu.

"I . . . am . . . Joe . . ."

"Jo." The teacher repeated the name of his first Occidental student.

That evening, Kid Greenstein found his way to a cheap inn where he spent the night, and early next morning took himself to Master Yamashita's training hall. When he arrived at the *dojo* the teacher supplied him with a hakama which, though freshly cleaned and starched, was so old from use that it was no longer black, but gray. Joe put it on, took his place at the end of the line, and followed the others, bowing on hands and knees to the shrine. While the others chanted softly in Japanese, he offered Hebrew morning prayers.

After the five minutes of meditation, there was a thorough warm-up with special attention paid to the joints. Training began with Joe being assigned to a partner, a young Japanese of slight build but with the stern expression of a Japanese mask. Joe did not know quite what to make of him. The first exercise was practice of alternately throwing and deflecting a straight punch, blocking inside at the forearm muscle. Punch . . . block . . . punch . . . block . . . punch. After fifteen minutes of pounding his arms against his partner's broomstick counterparts, Joe had grown a dozen swollen eggs on each arm, and could hardly stand the pain. The very expectation of another blow on his raw arms made him wince. If Kid Greenstein's undernourished partner was uncomfortable, he did not show it, and the implacable little fellow was starting to wear on him. As the pain went from bad to unbearable, the next exercise began.

The partners danced around the mat, pushing and pulling, trying to move each other off balance long enough to kick the legs out for a footsweep and a throw. His partner had already raised a few of those eggs on Joe's ankles, but Joe ignored them; this exercise was similar to the one he had first practiced with Volanko.

Joe could see Yamashita stolidly following him with his eyes. It was business hours, and the sensei's authoritarian manner had returned. It was time to show the stuff of which Kid Greenstein, American wrestler, was made. When his opponent tried the sweep and throw, Joe unbalanced his stance, snatched him off his feet, and slammed him to the mat. Perfect. Japanese Mask was stunned and shook his head. Taking a moment to be pleased with himself, Joe looked up to find Yamashita standing beside him. He did not offer congratulations, and Joe could see he wanted to speak but had no understandable words. He took Joe by the wrist and a disabling electric pain shot up his arm. He was helpless as the teacher walked him in four directions then tossed him to the mat and held him there.

"Okiro . . . okiro." Yamashita ordered him to get up, but he could not; the teacher held him pinned with only two fingers applied against the back of the elbow. When the Japanese allowed him up, Joe was unruffled and the demonstration had not been wasted on him. This treatment had been administered for his education, without malice. Yamashita made a soft downward movement with his hands, and Joe spoke the words for him. "Sensei . . . I . . . work . . . too . . . strong. . . . Technique . . . only . . . technique." Yamashita nodded; Joe had gotten the message.

Joe developed a profound respect for the Jujutsu teacher, an unusual and humble man who lived simply in two small rooms behind the school. His wife was a shy, somewhat homely woman who was only glimpsed as she went about her duties. On or off the mat, Yamashita was a quiet, spiritual man, who was always self-contained and even during the most rigorous training never raised his voice. He lived almost like a hermit, his entire life bound up in his students, and he would grow old with no thought of growing rich. There was no fraternization between Yamashita and his charges, but it was obvious that they were singularly devoted to him. It was the students who did light repairs, kept the premises swept, and unannounced one morning a week left groceries and small gifts at the door to his rooms.

Kid Greenstein attended two formal workouts a day, morning and night. At the end of each session he could hardly walk, and there was no part of his body that did not ache. After three days, he was one hematoma from top to bottom. Yamashita trained not students but survivors. This survivor was learning quickly, mating the Jujutsu to his Occidental wrestling skills.

"Dear Volanko," he thought, "If it wasn't for you . . . I would never get out of this place alive."

Each afternoon and evening, Joe limped to the marketplace or to one of the numerous outdoor food stands where he ate his meals. His breakfast was fruit, hot bean soup, and a bowl of coarse rice mixed with a raw egg and soy sauce. Yamashita gave him a blanket and allowed him to sleep on a mat in the back of the school. In return for his kindness and instruction, imitative of the other students, Joe brought vegetables and saké rice wine which he left at the teacher's door.

After two months he had acquired what he had set out to learn, and could stay no longer. Reluctantly, he informed Yamashita that he would be leaving the next day, and the teacher indicated his understanding without comment. After training the next morning, Joe returned the gray hakama and put his things together, and the teacher and students escorted him outside.

Now there was a kinship between Yamashita and himself, himself and the others. Joe Greenstein had arrived a foreigner, but he had proved as fanatic an athlete as they, demonstrated that he loved what they loved with a passion that transcended nationality or language. Even Japanese Mask, the little samurai, now considered Greenstein his favorite training partner. Joe had worked and paid for this acceptance; he had become one of them. All eleven members of the Yamashita Jujutsu School stood at the door and bade him a formal farewell as he presented the headmaster with a black silk robe sash that he had purchased in the marketplace, and thanked him for all his kindness. Yamashita offered nine words of English, each spoken with care as if it had been prepared for Joe's departure. Though the translator was obviously a local neophyte, the genuineness of the words was unmistakable.

"Jo . . . you . . . come . . . back . . . sometime," Yamashita said. "You . . . have . . . nice . . . spirit."

With his newly acquired skills Joe Greenstein returned to the docks where he boarded a ship for Galveston. His training in Europe and India, his wrestling in the United States, had all been polished with this study of Japanese martial arts. Though not yet twenty, he had already engaged in an international spectrum of self-defense training.

A few weeks later he disembarked in Galveston, and collecting his back pay the same day, sent a final letter to his wife.

Dearest Leah,

I'm sorry to have kept you waiting. I pray all is well, and that by the time you receive this letter our new child will have already been born. May he or she know a blessed and happy life. How hard it is for me to be apart from you both. Leah, it is my dream that we, all of us, will be good Americans. In this envelope you will find one hundred dollars. Come to Galveston now. . . .

Having returned to Texas in the summer of 1912, Kid Greenstein began looking for matches. Word came to him that Vladic Zybszco of the internationally famed and feared wrestling Zybszco Brothers was to have a handicap exhibition match with five opponents. Joe Greenstein volunteered to be one of them. On the night of the bout, Zybszco, who weighed well over 200 pounds, had made short work of the first four, when 145 pounds of Kid Greenstein climbed into the ring. Joe stared at the mountainous Zybszco; if he could just stay with the giant for thirty minutes, he would win a purse of fifty dollars.

For half an hour Kid Greenstein evaded, scrambled, taunted, and tricked, and as the bear tried to embrace the bee, the audience whistled and stomped its approval. In the end Joe left the arena with fifty dollars in his pocket and with this money as his grubstake, went looking for more matches.

It soon became apparent that Kid Greenstein was a little too good and could not be relied upon to lose even once in a while. In all of his matches he had never been pinned, and the outcome of any contest would probably be a foregone conclusion; therefore no betting. Many of the promoters passed him by, except for those who offered money for him to "lay down." "Go to hell," he told them. In a short time he had become the champion of as many southern cities as he had wrestled in, but his frustration was growing daily. Lightweight wrestling was not as popular as the heavyweight variety, but in the places where he did draw, often taking on opponents heavier than his weight class, purses and promoters had a habit of disappearing. Joe knew he was a poor self-promoter and a terrible businessman. With Leah and the baby arriving soon, he could not continue as he was, going from town to town, struggling for small-time matches that he knew he would win, and getting nowhere.

There was a man who had achieved the reputation Joe desired; George Bothner of New York City, the undefeated Lightweight Wrestling Champion of the World, was a living legend, a giant-killer who delighted in destroying heavyweights more than twice his own size. He had been world champion since 1902, and though in his mid-forties was still at the zenith of his prowess. In 1912 he had retired to dedicate himself to the training of others, but his return that year had been forced by public clamor for him to meet Henry Irslinger, the Middleweight Champion of England, who had decisively defeated the best of his American opponents. In their match, Irslinger was unsuccessful. There was no one to compare to George Bothner in reputation or ability. Kid Greenstein was as ready as he would ever be; he decided to challenge Bothner.

He sent a letter to Bothner, and while awaiting a reply returned to his job on the Galveston docks. His concern now was for a steady job to feed his family. Kid Greenstein's local popularity and his reputation as a good and reliable worker allowed him to approach some minor officials of local 310 of the longshoreman's union, and he found himself a union member. His earnings increased, supplemented by occasional local bouts.

For some time the pay boss of the docks, a man named McG—, had been extorting money from the nonunion men. In order to receive the brass-numbered check which was to be cashed in for a day's pay, one had to give McG— a dollar out of a three- or four-dollar weekly salary.

At lunchtime, Kid Greenstein sat in an empty boxcar with some nonunion men and watched with disgust as McG— entered. The payboss was a large man with a putty nose, well-dressed in a vested suit, who cleared his throat to announce his presence and demand payment, all the while holding the brass pay checks behind his back.

Reluctantly, one of the poor immigrants gave McG— a silver dollar. The man was giving away his child's dinner; it was too much for Joe to bear. He got up, cleared his own throat, and smacked the silver dollar out of the payboss's hand. Stalking out of the boxcar, Joe stormed into the office of Mr. Anderson, president of the Southern-Pacific Docks.

Joe told Anderson his story. Though disgusted at such goings on, Anderson offered a few cautionary words.

"Joe, you're a union man now. It's not your problem, unless you want it to be."

"Can't be helped," Joe said, knowing there might be trouble for him, as he signed the complaint papers in the office.

At this time Joe had a friend named Maurice who, like himself, was an immigrant, an easygoing sort who always had a laugh or a joke for everyone. Joe considered him a good friend, and they lived in the same rooming house, a kind of barracks for the working men. On occasion, Joe and several of the other men would wake up to find money missing from their pockets. "The bum's probably broke and hungry," Joe excused the thief to Maurice.

As the trouble with McG— reached the boiling point, Maurice came to see him. This was usual; he was teaching Maurice how to write English script, how to sign his name. Maurice put a blank sheet of paper down in front of him.

"Here, Joe. Show me. Sign your name."

Joe signed, and thought nothing more about it. Maurice thanked him and promptly left. The next day Maurice came to see him again. He discreetly looked around, then reached into his pocket.

"Here Joe, this is for you. . . ." He took out a small fortune in bills.

"What's this?" Joe laughed. "Where'd you get that kind of money?"

"I might as well tell you," Maurice explained, "because I don't want you to come looking for me." McG— had approached him knowing that he was Joe's friend. Together they came up with a plan to get the payboss off the hook. Maurice took the blank paper with Joe's signature on it, and McG— typed in a confession that the charges against him had been faked. For this favor Maurice extorted the sum of three hundred dollars.

"I knew you'd get mad"—Maurice attempted to disarm him—"so I thought I'd better give you this." He pressed half the money into Joe's hand.

"I trusted you!" Joe exploded, throwing the bills in Maurice's face and grabbing him by the throat, *"and you've ruined me."* Greenstein flung him across the barracks. Maurice jumped up and fled for his life.

Joe was desolate, having liked and trusted the man. Now he knew. . . . Maurice was the one who had robbed the workers, including himself, while they slept. Joe was totally disheartened; it galled him that he was such a poor judge of character. He was left with the $150 scattered all over the floor. He wanted to give it back, but to whom? To approach McG— and give him money would be as sure a sign of incrimination as to keep it. Joe's good name had been smeared and there might be trumped-up legal charges as well. He had to leave town. He sent the money to his sister Chani and went to Houston to start over.

In Houston, he got a job at the Samson Junk Company on Buffalo Avenue. In the daytime, he chopped scrap iron with a sledgehammer and scrubbed bottles. At night, he made his rounds as a watchman armed with a shotgun. Working night and day, he got eleven dollars a week and a place to sleep. Word was sent to Leah to meet him in Houston.

It was December, 1912, when the time finally came for Leah's arrival, and he dressed in his best and only suit, picked up the gift-wrapped blue dress he had bought for her after hours of deliberation, and climbed into the surrey. He had bought a horse and buggy at bargain prices and resurrected both. In his succession of owners, the old brown swayback had undoubtedly never had such treatment; Kid Greenstein brushed and groomed him as if he were a thoroughbred.

Joe waited anxiously for the Santa Fe train to arrive at Union Station, and when it did, he searched the crowd for a tense moment before he saw her.

"Leah!" he called across a maze of skimmer hats. She looked beautiful if a little disoriented, and carried year-old baby Judith in her arms. As he hurried toward them, she appeared suddenly unsure of herself, a shabbily dressed immigrant woman; the husband she had not seen in so long now appeared a man of wealth.

"Yosselle, you're a millionaire," she said.

"No, I'm a junkman," he replied and seized them in his arms. Leah knew then that it was her Yosselle, as always.

"How can a junkman own such a suit and a fancy buggy?" she asked. "How can a junkman bring such a present?"

"This is America." He smiled and held the baby and drove them home to the white frame house he had rented on Commerce Avenue.

Soon Kid Greenstein was broke again. After settling his family in, buying furniture, linen, food, and sundry extras for the house, his savings had evaporated. Leah didn't care; they were together at last.

Not long after Leah arrived she was pregnant again, and it was obvious that Joe's menial job at the junkyard would not pay their way. They relaxed on the creaky porch and Joe played his half-dollar Hohner harmonica as the sun set and wagons lumbered down the street in front of the house.

"We'll make do, Joe," she encouraged him, "not to worry."

A few weeks later a letter was forwarded to Joe, and when he saw the New York postmark he opened and read it, shouted, and ran out back to where his wife was hanging laundry.

"Leah, look . . ." He held the letter out to her.

"So what is it?" She continued with the wet wash.

"It's what I've hoped for."

"Dear Mr. Greenstein," the letter began, "I herewith accept your challenge. . . ." The letter was from George Bothner.

"Leah"—he sat her down—"when I beat George Bothner it will be the start of wonderful things. It will be different; we'll have what we need. You'll see. . . ."

Greenstein trained himself more mercilessly for this contest than for any he had ever had. It was early 1913 when Leah packed him a few meals in a paper bag and Joe took the railroad to Reading, Pennsylvania, where the match was scheduled. The train was delayed and he arrived the morning of the bout. At the appointed time he entered the ring through a packed local crowd. George Bothner stood in his corner. His face was pleasant and still boyish, rather handsome and without bravado. Joe liked him at first sight. Like himself, Bothner had begun his training as a puny and anemic youth. Now at forty-six, he was still sleek and wiry in physique; a human cat. The champion was a master craftsman of the ring, a scientist who knew it all. In 1904 in the Grand Central Palace in New York, Bothner had nailed Higashi, Japanese Jujutsu Champion, to the mat in three straight falls. Joe put it out of his mind. This was Kid Greenstein's moment. Nothing would change it.

The bout got under way and each alternated offense and defense, searching for a weakness. Bothner was a wizard. Toe to toe he was a tiger; hold him fast and an eel suddenly wriggled from the grasp; should Joe hold back to rest for a moment a spider of ten arms and legs enveloped him. Bothner was an athlete with a brain, one who measured applied energy with the astuteness of a physics professor. Joe tried to conquer him with speed and failed; Bothner stymied all attempts. Joe pulled out the stops and went after him with pure attack for fifteen minutes; Bothner sidestepped and evaded with the balance of a high-wire artist, and whatever Kid Greenstein attempted was short-circuited before he could execute his move.

Joe's every action was frustrated, and he began to wonder if his opponent was a mind reader. Bothner wasn't doing any better; when he went for the legs to upend his opponent, Joe caught him in a headlock that threatened to burst his head like a melon. Bothner took him down and attempted his favorite hold, the body scissors. Once it was applied there could be no escape; he could dominate any opponent with it. Joe barely slipped free and came at him again.

Using every pressure of technique, speed, and strength, each gave the other no quarter. Nor could either man pace himself; one who did not go all out might be overwhelmed in seconds. Each evaded holds that would have finished another competitor. They became mirror images of one another, thoughts and actions interchangeable; neither's skill could dominate the other, neither's spirit would break, neither would quit. After two hours and forty-seven minutes they had wrestled to no fall . . . a draw. When it was over both athletes were carried from the ring on stretchers.

Joe Greenstein sat on a bench in the railroad station physically and emotionally exhausted. How could he face Leah after all his grandiose talk? What was he to do, start the trail of small-time matches all over again? That he had wrestled to a draw the man whom sportswriters called "the greatest lightweight wrestler the Western world has ever produced" was of no consequence to him. He had not defeated the champion, which alone would have improved his position. He sat with his face in his hands when he heard a soft voice behind him.

"Hey, Kid . . ." He turned; it was Bothner. "I asked over at the hotel. They said you were probably over here waiting for the train," he explained. "I thought I'd come over and see you off."

"Kind of you, George." Greenstein managed a smile, and Bothner sat on the bench beside him.

"I don't know where you've been, Joe, but you haven't been wasting your time. Where'd you learn your stuff, anyway?"

"Oh, around," Joe said, too depressed to elaborate.

"You even studied some Jujutsu, didn't you?"

Joe nodded that he had.

"I figured." Bothner fell silent and a moment passed before he spoke again. "Joe, I just wanted to tell you that I threw everything I had at you today. You're a great wrestler . . . great."

"Coming from you, that's a high compliment."

"Well, anyway, you can tell your grandchildren that you drew with George Bothner," he said objectively.

"It's not enough, George," Kid Greenstein answered without conceit. "I came to win."

"Don't I know it, Kid. Don't I know it." Bothner rose. "Safe trip."

Joe Greenstein boarded the train with a sinking feeling, knowing that he would never meet Bothner again, that his great opportunity had slipped away.

Kid Greenstein returned to Houston and set up his own modest business, the Square Deal Junkyard. Word got around that his scales were honest, and business picked up. That a world-class wrestler had returned to the low estate of junkman depressed Joe, but he remained hopeful. He was making a spare living, laboring as a junkman in the daytime and wrestling locally at night.

With the birth of their son Harry, in 1914, Joe and Leah's roots sank deeper into the thriving and exciting city of Houston, which retained its Wild West atmosphere and where many people still wore guns as readily as socks.

One of the most eccentric of Joe's acquaintances was a young man by the name of Sam C., a moody sport given to erratic behavior and a fascination with firearms. But Sam was polite and always went out of his way to say good morning to Leah. Joe took little actual notice of him until one day Sam said, "Kid, sometime soon I'm going to take your wife." He smiled when he said it.

"Over my dead body." Joe laughed at him and continued on his way, knowing that he couldn't let Sam offend him; Sam wasn't quite right upstairs.

Joe was a close friend of the disturbed youth's father, and at the old man's request, stopped in at his business establishment, a shoe-repair shop at 2004½ Congress Avenue and Chartres Street. It was Monday evening, October 12, 1914. He walked inside, then heard someone call his name.

"Hey Joe!"

He turned. Sam C., a .38–.40 caliber revolver in his hand, fired from thirty feet away through the open door. The bullet slammed in between Joe's eyebrows. Knocked off his feet, he lay dazed, stunned. When he opened his eyes in shock, he felt the blood running down his face. He knew he'd been shot between the eyes. Not knowing what else to do, he absurdly put his thumb against the bullet hole to stem the flow of blood, rose shakily to his feet, and staggered across the street to a pharmacy.

Antwerp Needed This Man To Foil Germany's Cannon

Charging madly against the unsuspecting brow of Joseph Greenstein, 18 Chartres-st, a rampant bullet from a 32 caliber revolver came to sudden grief last night at ... ress-ave. easily picked it out from under the skin. The slight loss of blood was all the human target suffered.

"My friend was showing me his gun, and it went off," Greenstein explained. "I fell down. I did..." (a)

ARRESTS FOLLOW SHOOTING SAID TO BE ACCIDENTAL

Despite their protestations that the incident was purely accidental, B. H. and Sam Croftsic, shoemakers and father and son, are under arrest on a joint complaint out of Justice McDonald's court, charging them with assault to murder. The third person in the case, Joseph Greenstein of 18 Chartres street, is in St. Josephs Infirmary ... bruised forehead ... (b)

BULLET FLATTENED OUT AGAINST HIS FOREHEAD

Except for Shock and Slight Loss of Blood Joseph Greenstein Was Uninjured.

Shooting no longer has its terror for Joseph Greenstein, 18 Chartres street. Monday evening about 7 o'clock Greenstein was shot, accidentally, by a friend, the bullet striking him in the middle of the forehead. Instead of entering the skull, the bullet, which was from a 32 ... revolver, flattened out against ... (c)

(a) *Houston Press,* Tuesday, October 13, 1914; (b) *Houston Chronicle,* Tuesday, October 13, 1914; (c) *Houston Daily Post,* Tuesday, October 13, 1914

After several patrons had stampeded out, one fainting onto the sidewalk, the druggist stretched him out on the marble soda-fountain counter. He looked into Joe's eyes and felt for a pulse. The Texas druggist had seen gunshot wounds before.

"You alive?" he asked bewildered.

"Sure, I'm alive, schmuck. I'm talking to you, ain't I?" Greenstein replied indignantly.

In short order, an ambulance came and took him to St. Joseph's Hospital, where in the emergency room Dr. Louis Daly and several assistants removed the bullet.

In the hospital office, Joe sat with head swathed in a bandage. "Why am I alive, Doc?" he asked. Dr. Daly smiled with an expression of relief and puzzlement. "You're flesh and bone, just like everyone else. I can't explain it. Maybe you're a throwback, or a throw forward. Maybe it was a defective bullet. Maybe you're just the luckiest man alive."

Joe stood on shaky feet and buttoned his shirt.

"I've been a surgeon for many years, and I don't have an answer. One thing's for sure though"—clink, he tossed the small metal object onto the porcelain examination table—"you stopped that with your head."

Joe held the gray metal between thumb and forefinger . . . a slug . . . now as flat as a nickel.

Over the doctor's protests he took himself home, arriving to some local acclaim.

"KID GREENSTEIN STOPS BULLET!"

The newspapers had erroneously reported the bullet as a .32, but it had been in fact a heavier .38–.40, a .38 slug backed up by forty grains of black powder, a load first used in rifles, and well known to be a man killer.

That evening Sam C.'s father came to see him, hat in hand, and Leah showed him into the living room where Joe was seated in an armchair with his feet propped up.

"Joe." The old man wept at the sight of him. "How could he do it? How? If you want to get a rope and fix him yourself, I can't blame you . . . but he's sick. He's not right in the head. I swear I never thought he could do such a thing. . . ."

The old man had shown Joe many a kindness, and though he labored long and hard in his shoe-repair shop for every dime, he had offered to lend Joe money when Leah became pregnant again. The old immigrant's life had been bitter enough.

"Joe, I have no right to ask," he went on pitifully. "I don't know how you're sitting there. He killed you. I have no right to ask. . . ."

"What can I do?" Joe said.

"Spare my boy. I'll punish him myself. Don't do what he did to you. If you tell them he meant it, they'll put him away. You know what would happen to him in prison. They're animals in there. . . ."

"It was an accident." Joe cut him off in mid-sentence.

The man stopped speaking and wept in his hands. "You mean it Joe?"

"An accident. That's the way it was," he said again.

"Bless you. I won't forget it . . . never forget it." Leah ushered the old man out.

This magnanimity had not been so difficult for Joe to muster. He had no thought of revenge because he was much too contemplative about being alive. Sam C. never went to trial. Years later it was rumored this same young man ended his own life . . . with a self-inflicted bullet in the head.

Joe Greenstein had been preserved by something beyond understanding. This survival changed his life, his very perception of reality. He pondered the passage in the Hebrew "Wisdom of the Fathers," one of the treatises of the Talmud written between the latter half of the fifth and the third centuries B.C.E., which questions whether each man is master of his own fate, or if everything is predestined. Though opposed, both are true, the passage concluded. It is a paradox beyond man's understanding.

Volanko's admonition that nothing is impossible rang inside his head. Leah found him in the living room dressed in a robe and seated in his armchair, a motionless presence barely illuminated by a kerosene lamp.

"Joe, it's three in the morning. Come to bed. You need rest."

"I was on the ground. I knew I was shot . . . a burning pain between my eyes"—he lifted a hand to his head—"and then somehow I wasn't myself . . . but something much more.

"And I knew in that instant that I wouldn't die. Couldn't. Not even a bullet between the eyes could kill me. It's all been for a reason, Leah. Everything. Volanko . . . the bullet . . . everything. All for a purpose."

"What purpose, Joe?"

He looked blankly at her; he had no answer.

Joe Greenstein knew that his life had changed irrevocably; that nothing would ever be the same again. He took to clipping reports of others' flirtations with death. As he sat at a basement table with old bound newsprint stacked up before him, a librarian placed an open volume before him. "'. . . Mountain Climber in Perilous Fall.' This guy fell off a mountain, and held onto a rope in a snowstorm for four hours." Joe examined the report with interest.

"Where is that . . .?" The librarian flipped through another brown-edged volume. "Oh, here it is." He ran his finger across the tabloid page. " 'Woman Saves Own Child From Certain Death.' This kid was trapped under a milk wagon, and the hysterical mother lifted it and pulled the kid out. Amazin', huh?"

Indeed such accounts were amazing, but no more so than his own. Joe became convinced it was destiny that he live, yet he knew that he had somehow taken a hand in his own survival. There was more to it than mere chance. Now he collected these oddities of survival and supernormal achievement, searching for the key to why these particular people had survived. Bolstering his new consciousness, he purchased books on psychology, anatomy, and mesmerism.

He returned to serious training, running long miles each morning, lifting improvised weights at the junkyard, performing breathing exercises to condition himself internally. Several months after he had been shot, Joe Greenstein sat alone on the porch holding in his hands the spike which had become shiny from his constant handling, and he examined the object as if for the first time.

"Why have I failed to bend this? I know I am stronger than a piece of metal, yet raw strength always fails. I can feel I'm not using everything I've got." Then it came to him. "The *mind* holds back. Something within, an instinct for preservation, shuts down the system. An inner voice says, 'Do not go further or you'll hurt yourself.' "

But, if a man wanted to badly enough, could he learn to override the governor and use the mind to make the body obey and go forward past its learned limits? If he could cleanse all other thoughts and fears and use enough mental concentration, might he be able to focus the whole of himself through his hands?

"A life-and-death situation clears the mind of self-imposed limits," Joe theorized. "But if man has the capacity to do these things, why does he need an outside stimulus, a danger to make him react? Why couldn't the motivation come from within?"

With a new resolve he slowly, almost ceremoniously, wrapped the spike in a handkerchief, and with every wind of the thin cloth cushion, he divested himself of all anticipation of difficulty, fear, pain, or wish to go back. "I will bend it now. My mind will command my body. The spike will give way. . . ." And for the first time Joe Greenstein pitted himself unreservedly against the bit of iron and felt it give way in his hands. Excitedly, he unwrapped the cloth; the spike was slightly bent. He had come upon something quite remarkable. "It works. . . ." He attacked the spike again . . . again . . . until it was bent in half.

In the next days he began to experiment with techniques of optimum leverage. After a few weeks, he could bend a spike with four sustained attacks. Then he learned to do it with two swift motions. He demonstrated his skill to Leah by quickly bending the spike for her in the kitchen.

"Joe, that's wonderful," she said, "but what can you do with it?"

"I don't know . . . yet." It was the process that fascinated him, he had no idea what to do with such an ability. He did know he had discovered something within himself of real and far-reaching importance. "Imagine, a new kind of man," he thought to himself. "A man without learned limits . . ."

SAMSON REBORN

By the time Joe and Leah's third child Esther had been born in April, 1915, Joe had begun a transformation. He recalled Volanko's words: "I'm proud of you, Joseph, continue as you have . . . and someday you will be as great as Samson. Place no limits upon yourself and you will have none." The words now struck him as prophetic.

"Joe," Leah said, "you're so serious lately. If something's wrong, tell me. . . ."

"It's hard to describe," he answered, groping for a suitable explanation. "Leah, it's like when you sing so high and sweet; what if I were to try and catch the music in the air? I know it's there, I can hear and almost feel it. Still, it's invisible and I can't grasp it with my hands. But I think of touching that something that other people may not see, or know about, or even have a name for, but that I know is there." He smiled at her, thinking how foolish it all sounded. "You must think me crazy."

"I know you're not." She seemed to understand.

"Leah, I . . . I'm not like other people anymore." He made the awkward admission.

"Yosselle"—she took his hand—"you never were."

Joe felt the need to set himself apart from other men in

On their way to see D. W. Griffith's *Birth of a Nation,* Galveston, 1915

appearance as well as thought. He stopped going to the barber and let his hair grow long. It may have begun as a flight of ego or curiosity, but now it was an idée fixe, a promise of a new reality that beckoned from a far horizon. The eccentricity became identity; he began to think of himself as Samson reborn.

If Samson's power was in his hair, Kid Greenstein's locks would also be no mere adornment. Always one to believe in a scientific basis for the Biblical, he went to the library and read voluminously about hair, root, stalk, follicles; he concocted lanolin creams and oil compounds and massaged his scalp. He began gentle hair-pulling exercises as his hair grew longer, and lifted weights tied to his shoulder-length black mane. After all this effort, Samson's tresses would have to have a function. Other than knocking down a pagan temple, how would a modern Samson prove his powers to himself? He might pull something tied to his hair . . . an automobile, for instance.

Joe borrowed a Packard; the owner puzzled as he promised not to drive it. He set the heavy phaeton on a driveway facing uphill, secured a rope to its front bumper, and tied the other end to his hair. He tested the rope, then set all of his bodily power behind it. The rope snapped taut. He strained, faltered, then strained again. The car lurched forward. The pain of the unnatural feat brought tears running down his face, but the car was rolling uphill behind him. Step by step he towed it with his hair; he stopped after half a block.

He had done it. He had traumatized the tissue under his scalp. A bump appeared on the top of his head, rapidly growing from the size of a goose egg to that of a baseball. It looked horrendous, and hurt like hell. He knew Leah would get excited if she saw the gargantuan knot on his bean. That night he sat down to dinner wearing a hat. She immediately became suspicious.

"Joe," she inquired with her usual forbearance, "a hat at the table?"

"So?" he attempted to brazen it out.

"Joe, it's ninety-two degrees out. You're wearing a hat and an undershirt." He shrugged as if such dress were nothing out of the ordinary.

"Either you've become a Hasid," she said, referring to pious Jews who cover their heads at all times for religious observance, "or"—she reached over and yanked the hat off, staring incredulously at the bump.

"*Yosselle, what have you done to your head?*"

"I pulled a big car tied to my hair," he said with aplomb.

"Why?"

"Why not?" He tapped the hat back on and continued eating.

After several outbursts at dinner in which Leah pleaded for sanity, declaring that he had acquired another ability with no practical application, he agreed to let his hair take a holiday. He did so reluctantly, wondering what feats could be performed if he stayed with it. To his knowledge no one had ever done anything like this before. He decided to keep his hair long, and file the Samson routine away under future business.

By the time America entered the First World War in the spring of 1917, Mom and Pop Greenstein's fourth child Isaac was already a year old, and Leah was pregnant once again. Though mindful of his responsibilities, Joe was nevertheless caught up in the tide of nationalistic zeal that swept the country, and he had his own particular reasons. A Jew living in Eastern Europe was not truly a citizen of those countries, but an Israelite in bitter exile, oppressed by the local military. There, if possible, he would evade forced conscription, for servitude might last as long as twenty-five years. However, a Jew living in America was an American. Here, he would attempt to get *into* the army; Joe went off to sign up. The recruiting office greeted him with enthusiasm, until they discovered that he had a wife, four children, and one on the way. They explained that a two-cent German bullet might leave his wife and kids an expensive burden on the Government. Politely, they told him to go home.

Joe resolved to contribute to the war effort somehow, and if he could not help out with actual army service, he would find some other way. Having been shot, and disliking the experience immensely, he decided to save other American boys from suffering the same fate. He turned inventor. His contribution was "Greenstein's Bullet Deflector," a contraption intended to revolutionize trench warfare.

His invention was a conelike portable shield, made of heavy sheet steel; an American marksman could fire from the small opening at its center point, but an enemy would have a hellish time trying to shoot in; bullets would bounce off the steel funnel. "Greenstein's Bullet Deflector" even had a special place for mounting a sighting scope, and a rear-view mirror to detect any enemy attempting to infiltrate from behind.

He went to see his Uncle David.

"Uncle David, I want you to do me a favor."

"Shoot."

"Exactly," Joe replied, "I want you to get your rifle and take a few pot shots at a guy."

"What guy?"

"Me."

With the prototype of Greenstein's Bullet Deflector, his Uncle Silverstein, and a .22 caliber rifle in tow, he took an appointment with a Major General Spinks at Fort Crockett, Texas. With army brass looking on, Joe scurried up and down a trench while a previously reluctant Uncle Silverstein blasted away at him with the .22. The shots ricocheted off the Bullet Deflector, to Joe's elation.

The idea was turned down. "Impractical," they said. "A ricochet would certainly kill the man next to you. Besides, an army does not march into battle armed with iron funnels." This was his first and last run-in with the military. He was not about to give up, although slightly disheartened. How could he, a failed inventor, turned down for military service, contribute to his country in time of war? How could one man, with only his love of country and the strength of his arms and hair—?

The next week was a whirlwind of activity. There was no agency to approach for what Joe Greenstein had in mind, so he went about organizing it himself in piecemeal fashion. He contacted the Chief of the Houston Fire Department, who at first resisted but was shamed into acquiescence after an hour-long patriotic oration by Greenstein. Next, Joe went to Ellington Field, where, after similar blandishments, a cigar-chomping barnstormer agreed to supply his aged aircraft and equally shopworn services at a reduced rate. On Sunday, at the end of Joe's week of preparation, a dozen fire trucks were dispatched to the street corners of Houston, where from each truck the coming event was proclaimed:

"KID" GREENSTEIN RISKS DEATH TO DEFEND LIBERTY!
BUY WAR BONDS!

As the trucks carried their message to all on the ground, and the eyes of Houston looked skyward, Kid Greenstein, the hoary barnstormer, and his assistant took off from Ellington Field in a biplane.

Joe had it all figured. How much did he weigh? One hundred and forty-five pounds. It would be easy. Just take a little nerve, that's all. He would hang by his hair from an airplane. After pulling the car, Joe had developed a lock and combs for his hair. First, a metal plate slid over his long hair, then four combs inserted from different angles, and finally another metal plate on top. A steel-and-hair sandwich, the whole thing wrapped tight and tied to a rope. He had also brought along a belt of lead weights, which when fastened around his waist would help prevent him from getting blown around in the airwash.

Combs, hairlock, and lead belt in place, he looked down over Houston where the fire trucks were already starting to sell bonds. It was a long way down, but he would go through with it. The pilot leveled off speed, and Joe was lowered down on the rope.

Below, Leah made her way down the street pushing a baby carriage loaded with kids and bags of groceries, and she wondered why every eye on the avenue was turned toward the sky. She looked up and saw nothing more than azure blue and cotton clouds. There didn't happen to be a placarded fire truck nearby, and because Joe had not wanted to worry her he hadn't mentioned his scheme.

The drone of the biplane's motor became audible as the craft banked slightly for another pass over the city, this one right over the street where Leah was pushing her baby carriage.

"Here he comes again!" a man called out, as she caught sight of the plane and the curious individual who hung from it by his hair with hands casually folded in front of him. In that instant she knew. "Joe, my God!" When the shock of the sight had passed and the other women on the street had finished screaming, Leah collected herself and asked the usual question: "Why do that?" She knew Joe's inevitable answer: "Why not?" Reconciled to prematurely white hair, she continued down the street not daring to look up again.

But if Leah was terrified by the stunt taking place a thousand feet above, Joe was not. He couldn't really be very frightened because hanging by the hair prevented him from moving his head or looking down, and all he could do was stare straight ahead unable even to close his eyes because of the pressure that forced the eyelids up.

He had underestimated the buffeting of the wind. The weighted belt proved a hindrance and he flapped around in the air stream with every second subjecting him to stress that threatened to snap his neck. But the people below could see him, and they were forming lines to buy bonds.

Then it happened. Sheer terror. Another minor miscalculation. The hairlock and combs would work perfectly on a straight pull against an automobile, but wind currents and constant buffeting loosened the combs. The hairlock gave way. He seized the rope and yelled, "Pull me up!" but it was only inaudible mouthing under the roar of the motor. The pilot and his assistant couldn't hear a thing, and banked for another pass over Houston with Joe's frozen grip all that kept him from the long fall. The young assistant, wild with exhilaration at the speed and danger of the stunt, was so busy wahooing and enjoying himself that he forgot his duty as safety man.

"Pull me up, dammit!" Joe cursed at them and fought to hold with numb fingers. "Pull me up!" He was answered by another "wahoo" and a kaleidoscopic view of the treetops. Finally the assistant looked down. "G-Geez!" he stammered and hoisted Joe in. The plane winged back to Ellington Field.

That evening Joe sat at the dinner table wearing his now familiar hat, and ate quietly as if he had spent an uneventful day at the junkyard. Leah served him in stony silence.

"You'll bust if you don't talk," he goaded her, "so talk."

"You had to do that?" She vented her anger. "This was necessary? To get killed and leave four children and a pregnant wife. This is love, Joe?" She hurled his betrayal at him.

"Leah," he said quietly, "there are other men who love their families, but they are off fighting the war, and many will never come back. So, what will I do for America, nothing? No, that I can't do. You know how many Liberty Bonds I sold today? Plenty. Enough to make a real contribution. And if I ever ask myself, 'Joe, did you ever take a little risk for America,' in a small way I'll be able to say, 'Yes.' "

Clashes over Joe's recklessness became fewer as Leah reluctantly accepted the dangerous eccentricities of a man who made her feel loved and wanted, who was attentive to her needs though he could not lighten her burdens.

To Leah and Joe Greenstein, love, marriage, and children were the unchallenged natural order of things. On September 22, 1917, the natural order increased to five with the birth of daughter Minnie. Soon after, Leah was pregnant again.

"Joe, for cryin' out loud," his junkyard cronies would rib him, "didn't you ever hear of birth control?"

"Birth control? You don't understand," he would retort, "I love my wife."

"There is a great gift in the universe," he would say, "life itself." To Joe and Leah each new child was unique, born with infinite possibilities and capacity for creativity and love.

But for these blessings there was a price to pay. Joe's athletic ambitions had to take a back seat to his day-to-day responsibilities of raising a family. He had become a curious sort of man, an athlete/bookworm who still wrestled at every opportunity and in spare moments could be found haunting the shelves of the local public library. By now, his theories of untapped human capability, his interest in conquering inanimate objects by mind control, had taken him from iron spikes to thicker, more difficult objects—horseshoes.

Despite a lack of important matches, Joe Greenstein nevertheless constantly strove to improve his abilities. Even the most powerful wrestler still had his weak points—the groin, the eyes, the throat. For some time, he had been trying to overcome the problem of the choke hold, a threat to every wrestler. It was considered illegal, but not every referee could see it being administered. A headlock could become a choke simply by a deft change of pressure from the upper arm to the lower.

He prepared himself against the choke in several ways. Hunching the shoulders and pulling in the chin could in effect reduce the area of attack. As part of his training, he would hold his breath for long periods of time, sometimes for as long as three minutes. He had practiced this since his days in India with Volanko. It would give him time to evade or break a choke hold. But most important, he began to strengthen the muscles and tendons of the neck. When the chin was down and the neck tensed, these "cables" would form a natural armor to protect the trachea.

Arms laden with laundry, Leah entered her kitchen to find her husband in the center of the room, gripping the corner of a square oak table in his teeth. Its leg against his chest for balance, his eyes popping and veins in his neck and head bulging, he lifted the table and stood erect with it. Of all the exercises Leah had seen him perform, this was the most bizarre.

"What now, Yosselle?" she said mildly.

After doing multiple squats with the table clenched in his teeth, he set it down.

"I'm going to beat the choke hold."

"Oh," she said.

At the end of this development, the tendons of the neck were strong enough that when tensed they could withstand the blow of a fist. This also led to a rather remarkable discovery. There were teeth marks deep in the corner of the table. "You're ruining my table," Leah protested, so he cut a piece of galvanized tin and hammered it over the corner, then indented it to hold his bite. As he continued with these unorthodox exercises, he noticed tooth marks deep in the metal. He examined them curiously.

Human teeth couldn't cut metal . . . or could they? It seemed very far-fetched, but there it was. Maybe it was something about his teeth? He picked up a piece of wire, put the end between his teeth, and slowly bit down.

To his surprise, he could feel it give. He examined it; there was a dent. He bit again . . . deeper. He closed his jaws a third time and turned the metal with his hand. The wire cracked, bitten in two. He tried it again. It worked, and his teeth were none the worse for wear. He showed the feat to Leah. Her comment was succinct: "You'll break everything in your head."

He continued experimenting. He tried a bigger-gauge wire and halved it. As the gauge increased he learned how to apply the pressure, using a grinding motion, yet doing it with care. It was a combination of neck- and jawpower, naturally strong teeth and his new philosophy which excluded impossibility. After a year he could bite a quarter-inch-thick twenty-penny spike in half without injury.

■ ■ ■

Galveston, Texas: before departing for the oilfields

The junk business was a hard living at best. But now, Joe's expenses mounted while his small earnings decreased. He had to feed his large family growing by one child per year, and responsibility demanded that he act. Word was around that there was a new oilfield, a gusher not far from Burkburnett, a place called Newtown. Having worked for Benny Biskin, he knew something about the mercantile trade, and decided to try his hand at it. Using his modest savings to prepare and outfit himself and the family, he packed everyone into his Model T, whose body, from the dashboard back, he had hand-built of sheet metal into a light delivery van, and they took off for the oilfields.

They arrived in Newtown, which was not actually a town but a single rude street in the midst of an endless flat of mud and drilling rigs. There he set to work, singlehandedly building a store, house, and sidewalk. When he finished, he hung up his shingle: SQUARE DEAL DRY GOODS—J. L. GREENSTEIN, PROP. The Square Deal sold coveralls, boots, slickers, ammunition, anything one needed to survive in the oilfields.

Newtown's only hotel had a dance hall downstairs and rooms upstairs. On Sundays, the hotel was converted to use as a church, and the rest of the week it served as a whorehouse. Violence in the streets was not uncommon. A few months after they arrived Leah was out back washing clothes over a tub, while a few feet away Harry played and Judith stood minding baby Mary. A shot rang out and whistled past Leah's head, blasting a chunk out of the back wall. "Children, go play inside. . . ." She took casual notice of it, for wild shots were common, and went on with her washing. On one occasion the children were locked in the house because a lynching was going on outside, routine justice for a gambling murder. Newtown was a place for hardy souls.

For the Greenstein children life in the oilfields was a Huckleberry Finn adventure, whatever kind of grim last-chance gamble it might be for their elders. One never had to be alone because brothers and sisters were also best friends. When it rained their pop's store was a playland, and when the sun shone the nearby grassland beckoned them to run and explore. All things considered, after ten months, the family had much to be thankful for. Family affairs were going well, and Leah was pregnant as usual. The Square Deal had proved a success. Joe had, in fact, already made a small fortune which he reinvested in bigger and better inventory.

The family was asleep in the house behind the store when they were awakened by three quick shots out in the street—the warning alarm. A hotel patron had murdered one of the prostitutes and to cover his crime had set fire to her room. Another of the girls rushed into the flames after her and never came out. In a few minutes the whole of Newtown was a howling bonfire.

"Leah"—a startled Joe shook her awake—"smoke." She opened her eyes as a torrent of flames swept over their roof. He quickly threw blankets over the kids and carried them out two and three at a time by arms and legs.

At night he had made a practice of removing engine parts from his Model T and locking a long chain through its wheels to discourage theft. The vehicle now stood beside his burning building, and if it too went ablaze, the family would not only be ruined, but stranded as well. Though all four wheels were locked tight, Joe pushed the car sled-style through the mud away from the burning house.

While he was occupied with the car, Leah, though heavily pregnant, rushed back into the house, seized a pillow from the smoldering bed, put it over her face, and stood on a fiery chair to retrieve the family photos from the mantle. The Greensteins came out of the Newtown fire with only the scorched photographs, Joe's box camera, the car, and the hundred dollars in Joe's pocket. Looking around, Joe observed that they had been more fortunate than most. One old man, burnt out of a lifetime of savings and toil, had lost his mind and sat crying in the mud. The residents of Newtown milled about, dazed, faces blackened, many having lost their dreams along with their possessions.

Joe packed the family into the car and drove through the night to Kemp, Monger, Allen, an oilfield-company encampment some miles away. With seventy of his last dollars he bought a used tent, which he pitched in the oilfield.

"I thought we really had it made there for a while," he said, as the beat of the rain on the tent finally lulled the kids off to sleep. He and Leah were soaking wet, their faces blackened with soot like a pair of variety comedians. "Hell, I guess I make a lousy merchant prince. What do you think, Leah? Back to Houston?"

"No"—she shook her head—"we'll stick it out." He kissed her and they slipped into an exhausted sleep, their bed a blanket on the wet ground.

They started over, living in the tent. For Leah, these were the toughest of times. If her children needed bread, she had to bake it. Everything that was needed had to be made with one's own hands. Though Leah never complained, Joe swore to himself that someday he would make it up to her. The children seemed to adapt to the harsh conditions and the abrupt change in their lives, but though they played as before and appeared happy, Joe knew the situation was becoming desperate. Often there was not enough to eat.

He built another store, but with few goods to sell, and those ordered on credit, there was little income. He decided to let Leah run the new Square Deal and went off to get a job. The Macy Drilling

Company was a wildcat operation which everyone said would have oil blowing out of the ground in short order; for Kid Greenstein, it was the place to go. He took a horseshoe from his scant wares and searched out the rig foreman, a man of forbiddingly torpid gaze.

"You the boss?" Joe inquired.

"That's right."

"You do the hiring?"

"That's right," he repeated monotonously, still half ignoring the little man in front of him.

"I want a job," Joe said flatly.

"What can you do?" the man asked, though with little interest.

Joe raised the horseshoe chest-high and twisted it in half; then he handed it over for perusal. "Let's just say I'm a hard worker."

They hired him. He was a roughneck, which meant tough, filthy, and sometimes dangerous work. At night Joe would return to the tent covered with black grime, resembling Uncle Remus's tar baby. He would pass out on a cot only to awaken at dawn, and playing his harmonica to keep up his spirits, make his way back toward the obelisks that stood darkly on the mud flats.

Kemp, Monger, Allen, better known as KMA, was little more than a shanty town, but everyone knew everyone else. One of the good citizens of the place was a man called Clarkie, a skinny and mustachioed pimp whose "business" was a pair of sporting ladies. He would buy in the Square Deal, his purchases usually confined to dice or ladies' underwear. While Greenstein had no use for the man morally or professionally, he modified his disapproval because Clarkie was the kind to help you out if he could.

"Joe," he promised, "when your woman's time comes . . . you got reliable transportation." The mud in KMA was oftentimes over the knee, but Clarkie's automobile, a yellow 1918 Holmes, would go over or through anything. A terrific car. Even then, pimps drove flashy wheels.

When Leah felt the pains of her seventh child, she rustled Joe out of bed. He went out and found Clarkie, who, stark naked and somewhat drunk, was in the process of breaking several commandments at once with a woman he reluctantly ushered from his tent. A man of his word, Clarkie quickly sobered; they cranked up the

Holmes and broadslid through the bog toward Holliday, the nearest town where a doctor could be found.

It was pitch black as they walked up the steps of the doctor's house and pounded on his door. After a moment a light came on, and a profane old man stepped out onto the porch with his suspenders flapping.

"Who the hell is that? Who's botherin' my sleep?"

"My wife needs you. She's having a baby," Joe said.

"I'm not goin' out into the night to who knows where, with two jokers I never seen before," he said flatly.

"She's too far gone," Joe explained. "We couldn't bring her."

"Not goin'." The doctor closed the discussion.

Before Joe could argue further, Clarkie had reached under his coat and produced a mammoth horsepistol. The blued blunderbuss was more cannon than sidearm; he drew back the hammer and pointed the barrel squarely at the medical man's chest.

"*Either you're goin', Bones . . . or I'll drop you right where you stand.*" One peek at Clarkie's depraved mug and he surmised it was no idle threat. So back they went through the night, the doctor, the pimp, and the wrestler; back to KMA where Joe and Leah's seventh child was born in the oilfields on December 6, 1920. The good doctor evened the score by neglecting to register the birth. Mike Greenstein finally received a birth certificate at thirteen years of age.

By the New Year an association had been made between the Square Deal and the Macy Drilling Company. The outfit needed supplies and Greenstein sold them what they required, although heavily on credit. Everyone knew that there would be a big strike any day, and when the gusher came in, all accounts would be settled. But it never came in. The Macy Drilling Company went under, owing Kid Greenstein $900 for goods delivered, and $800 in back work pay. Now after two years in the oilfields, having used his savings to chase the rainbow, he had played out his hand.

"What do you think, Leah?" he asked.

"We've had enough, Joe. Let's go home," she said. They packed seven kids and his box camera into the Model T and never looked back. Though it was only about four hundred miles from KMA to Houston, the drive took nine days, often through axle-deep mud.

They took turns walking ahead of the car after they had nearly driven into a ravine where the road abruptly ended. There were few actual roads and no road numbers; the way to go was simply marked KT—King's Trail. For their dinner, they stopped by a lake and Joe shot some gamebirds and a few photographs.

Upon their return, Joe began peddling ice cream. Leah, always an excellent cook, took people into her Sunday table, serving a full-course dinner for thirty-five cents a head. They had hit bottom. And though all of Joe Greenstein's money had been lost, all of his plans frustrated or gone awry, though he, at times, felt like conceding defeat, he could not. The little voice was still there, and it said: "Be patient. Nothing is wasted. Everything in its time."

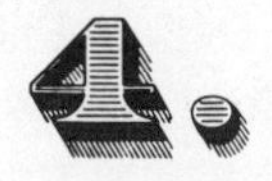

ODIN AVENUE

Kid Greenstein had learned to drive in 1911. A self-taught mechanic, he had always repaired his cars, mostly because he was usually too poor to pay someone else to do it, but also because he liked the work and possessed mechanical ability. He had always toyed with the idea of going into the car business.

He therefore listened with interest when a friend by the name of Sholom Kaplan offered him a new enterprise—a gas station. The place at 2310 Odin Avenue near the railroad tracks had fallen into disrepair when it had been vacated by the previous tenant, who had set up a garage across the street. Kaplan saw it as a mutually beneficial arrangement for Kid Greenstein to take it over and offered him a low rental and a small loan with which to get the place into shape. The deal was made.

Joe set to work, scrimping all the way. He dug a deep pit and set in underground gas tanks, installed a used vulcanizer for the repair of flat tires, and set up shop with the tools at hand. He put down a new cement sidewalk.

When he was finished he dubbed the place "Your Filling Station," and his new endeavor was launched. The neighborhood of Odin Avenue was in Houston's Fifth Ward, where Joe's patrons were almost exclusively poor people. Soon word spread about this shoestring operation and the wizardry of its proprietor.

One did not bring an immaculate Packard here; one pushed in his rattledebang Dodge with whacked-out valves. Once in safely, it was almost guaranteed that one would roll out under one's own power. What Joe didn't know he learned on the spot, tearing into the machinery with a vengeance and a greasy wrench in his hand. It was a family business; the older kids pitched in, dispensing gas from the two hand pumps, making change, patching wounded tubes. Mechanical endeavors were given a woman's touch; Joe taught Leah how to grind valves.

The family lived on the premises, which consisted of a large house partially converted into a filling station. There was a porch in the front, shaded by a thirty-foot pecan tree, and one in the back, a shop, a place to retread and vulcanize tires, and a watermelon stand where whole melons sold for a nickel. A fig tree stood at the left, in front of the place. Inside the house fabulous aromas from the Peerless Bakery down the street wafted through the open windows. The living room was decorated in early gymnasium, sans furniture, its floor wall-to-wall wrestling mat. Strewn about were lifting weights, Indian clubs, a medicine ball, jump rope, and the other accouterments of speed and fitness. There at odd moments Joe would tussle with his kids seven at a time; and on Sundays his ring cronies would stop by and the walls would shake from the sound of their combat.

With so many mouths to feed, Joe could not spend money on such frivolities as cast-iron lifting weights, which at the time were quite expensive. In Europe he had improvised by tying sand-filled sacks to either end of a heavy pole. In his junkyard days, he would stay late after business hours, lifting the forty-pound scale weights in each hand. Now, behind Your Filling Station, he made his own apparatus by taking an iron bar and setting either end inside a box mold into which he poured concrete. When the concrete set, he had his weights, which he made in various configurations.

On winter nights, when the temperature dropped below freezing, they would sit in front of the fireplace with Mom in one rocking chair, Pop in the other, and the kids on pillows warm by the fire. Leah would sing in her clear soprano voice while Joe played the harmonica and the kids sang along. It would have been a family

scene right out of the Old West, except the songs they sang were in Yiddish, the songs of European Jewry which in scarcely twenty years would be faced with extinction. The Greenstein children were fortunate to have learned their Yiddish in Houston.

At times Your Filling Station resembled an animal shelter as much as a garage. Now that life was more stable and quarters afforded more space, a menagerie of strays moved in with them. Neither Mom nor Pop Greenstein could see man or beast suffer, and none was ever turned away. On hand were always at least two salvaged dogs, a nest of cats, a hutch of rabbits and sundry other discarded furries, many being nursed back to health. Out back was a Holstein cow named Daisy, a Jersey calf named Baby, and a coop that soon multiplied to one hundred head of white leghorn chickens. The kids ran around the place barefoot. All in all, Your Filling Station looked like a Jewish "Tortilla Flat."

The Greenstein household was also a place for stray people. Your Filling Station was situated near the railroad tracks, and it soon became common knowledge amongst the down-and-out that the station's owners intentionally left their door unlocked. Regularly, one or two Knights of the Road would, after endless hungry days on the rods, jump their freight, walk over to the station, and bunk down for the night on the mat floor of the living-room gym where Leah always left a few neatly folded blankets for their use. The kids soon became accustomed to stepping around a snoring "guest" on the way to the kitchen in the morning. With a hot breakfast and a little friendly conversation if he wished, a drifter could go on his way somewhat renewed. The Greenstein family found that at one time many of them had been men of substance; they were often interesting and intelligent men who did not easily forget a kindness.

When neighbors suggested that Leah and Joe were taking a chance, that a hobo might steal, or harm the children, or even kill everyone in their sleep, Leah rebutted this as nonsense. "Just because a man is down on his luck, just because he looks like a bum . . . that doesn't mean he hasn't got a heart, doesn't mean he's not a person. A safe place to sleep, a hot meal, a little trust—what does it cost?" True to her words, not one of their guests ever made them regret their kindness.

And these men, whose luck had failed them, who had nothing worldly to give in return, made their feelings known. One morning the family awakened to find that a departed house guest, obviously a onetime artist of some ability, had left a gift behind: a rose meticulously drawn on a corner of the kitchen floor. A gangly cowboy from Oklahoma taught Judith some hobo craft, how to fold candy-bar wrappers into drinking cups.

Though Joe and Leah never meddled in their guests' lives, they did find themselves caring about them. Two blond young men, boys really, well-mannered and polite, had spent the night and stayed for breakfast; and when Leah gave them a sack of sandwiches for the road, they were all waves and smiles as they said good-bye. A few weeks later word came back from other men of the road that farther down the line, the two had fallen asleep in the hayloft of a barn, and when it caught fire they were trapped and burned to death. Hearing of it, Leah cried the whole day, though she never knew the boys other than by their nicknames.

Joe Greenstein had a way with kids and animals and practiced his naturopathic cures on all comers. With so many mischievous kids darting about, there was an endless parade of patients, and Joe always spent part of his day ministering to their wounds which ranged from imaginary splinters to possible concussions. There was no better and more trustworthy a healer than Pop Greenstein, and no more tender a bedside manner. In spare moments, one could find him reading one of his books on anatomy or herbology in the parlor or at his secret hideaway, the medical section of the Houston Public Library.

That summer, his amateur attempts at healing went professional. Leah became concerned when she noticed that Judith was walking with a limp. The child had been bitten by an insect, and the bite had festered. She had said nothing of it, and when Joe finally saw it, he knew it was a full-blown emergency. He rushed her to the doctors, but even after treatment, it did not improve. The infection had gone too far.

At last, the doctor drew his bottom line. "Mr. Greenstein, your daughter was brought to me too late. She'll lose her life unless . . ."

"Unless what?"

"Unless the leg is amputated."

In the next room, he could hear her crying, "Daddy, are they going to hurt me?"

He went out to her. "Sweetie, I won't let them. Now listen to me." He smiled at her while speaking softly, and she stopped crying. "Years ago, when I was about your age, three doctors said I was going to die, *and here I am.*" He beamed. "Just because they say something doesn't make it so. Now, your leg is sick, but we can fix it if we want to badly enough. Just you and me. And when you're all well, I'll buy you real silk stockings, and the prettiest dancing shoes in Texas, okay?"

"Okay, let's fix it, Pop!" she said.

"That's my girl." He lifted her in his arms and kissed her.

"Mr. Greenstein . . . ," the doctor said privately in an excited whisper, "without surgery this child will die."

"She'll make it. I'll see to it."

"Sir, if you can save that child's leg . . . I'll eat my hat."

Joe carried his daughter out of the office.

Discontinuing all else, he lanced and drained the swelling, then set about treatment with a constant washing of boric acid and application of his own concoction of medicinal and herbal powders, including: iodoform, seraform, bismuth, and later burnt alum. The leg was washed and lightly powdered every three hours, the bone-deep wound left to heal in the open air. He buoyed her spirits: they sang and told each other jokes; he made up stories, always with a happy ending.

They kept the vigil together for twenty-one days. At last, the infection abated, the leg began to heal. The wound left a tiny scar which was quickly forgotten as Kid Greenstein marched his daughter into a fancy Houston store. He slammed his palm down on the counter for service. "I want real silk stockings and the prettiest dancing shoes in Texas!" Then they went home, and he got his good felt hat, brushed it, and tapped it on his head.

He did not stop ringing the doorbell until the doctor answered. The man looked quizzically at him as he wordlessly gestured to the smiling little girl, who excitedly wiggled her toes inside squeaky new shoes. Abruptly, Kid Greenstein whipped off his hat and jammed it under the doctor's face.

"You want salt or pepper on it?"

The offer was declined.

Joe and Leah had been attending night school in Houston, and by the spring were ready to try for their formal citizenship papers. They appeared before a Judge Atcheson, and after answering several questions of the inquiry, Greenstein presented the judge with a poem he had written:

AMERICA! I SING FOR THEE
by Joseph L. Greenstein

Among its verses were the following:

America! I sing for thee and thy liberty . . .

The Home of the Brave, and the Land of the Free,
America, thou shall never cease to be,

Though I am not a native born, yet from my heart,
And from my soul, all other flags are torn, . . .

Never perish from God's earth shall be this
blessed Nation,

> Because it has given many homeless men a free
> standing station,
>
> A president is good for us, we do not need a king,
> England had a bunch of them and couldn't do a
> thing. . . .

"You wrote this?" the judge asked.

Greenstein nodded.

"All thirteen verses?"

"Yessir. Thirteen verses, thirteen colonies."

Atcheson adjusted the glasses on the bridge of his nose. "Greenstein . . . aren't you '*Kid*' Greenstein, the wrestler?"

"I am."

The judge began signing his name to the papers. "I don't think there's need for any further examination. Any man who could write a poem like that is already an American . . . and so is any woman he might marry." Joe and Leah Greenstein received their papers of naturalization on April 15, 1926. In a fitting celebration of their citizenship, Leah gave birth to their ninth child just before the New Year, a girl they named Rebecca, dubbed Rivie by the other children.

If Judge Atcheson had pronounced Joe and Leah to be full and equal Americans, there were those who differed. By 1924 the national membership of the Ku Klux Klan had swelled to over four million. Activity in Texas rose, with white-sheeted night stalkers lynching blacks and fomenting violence. It disgusted the newly naturalized Joe that the yahoos of the Klan should presume to define who and what was American. He had lived among ignorant rabble in Europe, and the memory was still fresh in his mind. He spoke against the Klan to anyone who would listen.

His words did not go unheard. Disturbing incidents began, small at first, usually confined to a little nighttime vandalism. He dismissed these until he began to find notes on the garage door: "WE ARE WATCHING—KKK." Several such notes appeared during the night, but still he continued to vilify the Klan.

In the afternoon, as he was taking apart an overworked engine, a car rumbled by and a tossed rock shattered his station window. With a quick sprint Kid Greenstein was across the station and up on the car's passenger-side running board. Hanging on with one hand, he struggled to unlatch the door and get in at its occupants as the car picked up speed. After two blocks of a wild ride, he was finally hit across the head with a lead pipe and fell from the speeding vehicle into the dirt. He lay stunned, but raised his head to see the receding license plate. Having memorized the number, he brushed himself off and returned home.

The next day, he had the plate number traced by a friend at city hall. It turned out to be registered to, of all people, the previous tenant of Your Filling Station. The man who was now his competition across the street was a Klansman. That was how they knew so well what he was doing and saying.

His first impulse was to go over and break the man's back, but they were probably expecting him. In Texas, if a man's premises or his home was invaded, he could kill the intruder and go free. Protecting one's property was a "mitigating circumstance." Kid Greenstein would have to bide his time.

A few weeks later, after closing up shop at one A.M., he went around back to check on his cow Daisy. At once, six men in white hoods stepped out of the darkness, several carrying drawn revolvers, hammers pulled back ready to fire. He dare not even struggle as he was bound and a gunny sack thrown over his head.

Shoved roughly into the back of a beat-up sedan, he was taken along with one other captive with whom the Klan had a score to settle. The car hurtled for an hour through the Texas night. "C'mon, fellas," he taunted them through the sack, "untie me, and put those pop-guns away. We can settle this." Silence. "What's the matter boys . . . no guts?" He was answered by the butt of a pistol slamming into the side of his head. When the car stopped, both victims were dragged out and their blindfolds removed. They narrowed their eyes, adjusting to the barrage of light and heat; a large, burning cross marked the land before them. A score of men robed in white surrounded them wordlessly, and then the beating began. With pipes and truncheons, the Klansmen fell upon their bound victims. After five minutes, Joe lost consciousness.

The sun was just coming up as he came to, lying naked beside the road. Blood ran from his open cuts; he was too weak to stand up. He had been tarred and feathered. A passing milkman out on his morning route stopped and picked him up. He never found out what happened to the other victim.

Leah had spent a sleepless night walking the floor with a colicky baby, and anxious over Joe's absence. When the milkman half carried him through the door, she somehow kept her composure at the sight of him.

"The Klan . . . ," he said before she could ask, and gritting his teeth, lay down on a couch, an agony against his flayed skin.

"Joe, how do I . . . ?"

"Kerosene," he said.

She bit her lip and cleaned his blistered skin, which came off in patches.

"It's not . . . Europe." He writhed at every touch of the solvent-soaked cloth. "These bastards won't run me off."

"Joe, would they hurt the children?"

"I don't know . . ."

"We'll sell the station, and leave here . . ."

"No." He would hear none of it. "Leah, do we have a right to live?"

"Yes," she said. "We have a right." Before he healed he was back at work. And still he would not keep silent.

Next to Your Filling Station was Foreman's Used Books Store, where Joe would pour through the musty shelves in search of this bit of minutiae or that obscure fact. Buying books was his pleasure, and Foreman's was his favorite place to do it. He had just purchased a huge and splendid dictionary for twenty-five cents when Mr. Foreman sidled over with a troubled look and whispered, "There's two men waiting outside, Joe. They look bad. . . ."

Sure enough, two bruisers with hands in pockets were loitering outside, trying to act unobtrusive. By the way they stood, they appeared to have weapons in their pockets. Joe immediately recognized them as two of those from the speeding car whose license he had traced, and no doubt they had attended that evening "service."

He thanked his friend Foreman and with the dictionary under his arm ducked out the back door, staying low to the hedges. He scrambled into his place through the rear, went right to the shop, and grabbed the long butcher knife that he used to cut watermelons. After pausing to sharpen the blade to a razor edge, he walked out into the street.

The two Klansmen were still facing the book shop, waiting for him to come out unsuspecting. Instead, they turned at a voice behind them.

"Hey! You rednecks looking for me?"

They stared at him.

"Come on!"—he invited them over—*"and I'll cut your throats!"* They froze, declining the offer. He advanced toward them. A professional wrestler with a sharpened butcher knife was not the kind of thing to get involved with before dinner, and though armed, the two beat a hasty retreat.

That night, Joe Greenstein sat down to dinner and offered the blessing over the meal. When he had finished, he nodded his satisfaction and added, "See, children, just like I told you. If you get them down to one or two . . . they're gutless." He smiled through split lips.

At this time, Joe's best friend was a young man named Benny Proler, an immigrant boy who had come from Lancaster, Pennsylvania, to Houston. Benny's atrocious English camouflaged a sharp mind. He operated Proler's Junkyard, and supplied Joe with everything from fenders to entire engines for rebuilding.

There was a new show at the Majestic Theatre, the headlining act a well-known vaudeville strongman, the Great Kronas—"The Man Who Sleeps on a Thousand Nails." Kid Greenstein got himself dressed up, and the two friends went to the matinee. Joe could hardly wait through the various singing newsboys, dancers, and bird acts until the Great Kronas came on. Finally, the time had arrived and Kronas went into his act. Kid Greenstein was impressed, never having seen a professional strongman perform in a theater. For the finale of his act, Kronas's master of ceremonies called ten men on stage from the audience; one of the volunteers was Joe Greenstein.

Kronas then went through the standard disclaimer, "Gentlemen, have we ever met before? . . ." He produced a formidable bar of steel which he was to wrap around his arm. Each of the volunteers was first presented with the bar to test its strength. When the bar got to Greenstein, he put it over his thigh, tensed, and it gave. He was about to finish it off when, "Thank you, sir," Kronas said, abruptly seizing the bar and holding it up. The MC went into the finale. "Each one of these men, the pride of Houston, have tried and failed. Only the Great Kronas . . ."

"Excuse me," Joe said shyly. "I think I can bend it."

Kronas turned and looked at him, at once sizing him up; his hardened extremities all adding up to trouble.

"I can bend it," Joe said again.

"*Then go home and do it,*" Kronas advised under his breath and proceeded to finish his act.

Joe was grossly insulted. The man had no business calling him onstage if he was not going to offer a fair chance.

"Forget it, Joe. Don't make such a big deal about it," said Benny Proler playing the diplomat.

The more Joe thought about it, the more infuriated he became, until finally, he knew what he had to do. They went to Benny's junkyard and picked out a bunch of steel bars the same size as the ones that Kronas had mangled. Then, Joe called up the Houston newspapers and gave them the scoop. "Kid Greenstein is going to duplicate the act of the Great Kronas." But whereas Kronas had done his performance before a packed house at the Majestic, Greenstein was going to do his in front of Your Filling Station. George Kronas was a well-known strongman, and popular on the vaudeville circuit. He was no phony, and for a local boy to challenge him would be at best foolhardy, and at worst a coronary.

Benny Proler was unsure. "Joe, how do you know you can do all that stuff?"

"I know." Greenstein was not to be dissuaded.

The Houston papers dispatched their reporters to the Greensteins' garage. There, sweating over the iron bars for the better part of an hour, Joe proceeded to duplicate the act of the Great Kronas. That evening, articles appeared in the newspapers, headlined:

HOUSTON VEST-POCKET STRONGMAN HAS BEEN FOUND

"Hot stuff, Joe," Benny said, examining the papers with enthusiasm, as Leah sat knitting. She made every attempt to avoid getting involved in such egotistical craziness. Joe was unimpressed by his clippings, having only imitated Kronas. "Now, I'll finish that guy. I'll do one for the papers, and if he tries to do it, he'll pull his fool head off. . . ."

"Joe." Leah dropped a stitch knowing what he was up to. "I'd better get your hat cleaned."

The Samson stunt. Joe Greenstein could not be content with pulling just one car this time. He decided to attempt three at once.

To prepare himself for the auto-pulling stunt, he resumed his massages, began lifting weights tied to his hair, and developed yet another outlandish practice. He placed the top of his head against a wall and pushed, applying pressure evenly, a kind of rolling pin over the scalp. Obviously an exercise for a madman, but it seemed to work.

Ordinarily, the hair follicles appear like bulbs under the scalp. He theorized that if they could be "exercised," they would expand and act as a kind of lock which would not pull out under tension. After a time, he could feel a thickening of tissue, a "muscle" under the thin skin of the scalp. Part of it, no doubt, was healed scar tissue.

Benny Proler was the type who could get things done on a moment's notice; when Joe called and said he was ready, Benny talked a local Stutz dealer into supplying the cars and informed the newspapers. This stunt wasn't going to go wrong like Joe's hanging from the airplane. The way he figured it, three cars would be a little more difficult than one. If there was a variance in the weights of the vehicles, he would place the heaviest one first and the lightest one last. He would keep the vehicles close together, but leave a few feet of slack in the rope so that the momentum of the first car would help to jar the second forward. The movement of the first two would pull the third. He would have to expend maximum energy until the chain reaction was completed, and then keep the motion going.

That Sunday, Kid Greenstein and assistant Benny Proler took their positions on a quiet street with little traffic and few bystanders. The newsmen were present and so was the motion-picture-news cameraman who had filmed his duplication of Kronas's act. The reporters set up their still cameras, fully expecting the first shot to be that of a man with a broken neck.

Greenstein set his hair locks and combs in place and tied the rig to a rope which had been fastened to the lead Stutz's front bumper. He would pull the three cars while facing them and walking backward. He found this to be the optimum position.

He set himself and tested the first car; the Stutz didn't budge. He told Benny to check the handbrakes of the vehicles and make sure they were out of gear. He knew from his previous experience that he would have to endure piercing pain and overcome it.

He psyched himself for the attempt, and with that strong wrestler's stance moved backward, as the white pain flashed in front of his eyes. The first car was rolling. He persisted, another step . . . another . . . the vehicle was gathering momentum. Now, when he would succeed or fail, he ignored the pain. He put everything he had behind the trailing rope, which snapped taut. The second Stutz jerked forward and began rolling. Sure enough, the first car's momentum had helped. With the two cars moving, the second trailing rope sprang up and the third Stutz lurched forward.

As the cameras clicked and whirled, he moved back step by step, the three vehicles lashed to his hair. After a half block, he turned and posed for the stunned reporters. Mission accomplished, he returned home with a headache, but satisfied that he had finally one-upped the Great Kronas.

"Leah"—he kissed both of her cheeks—"where's my hat, sweetie?"

"My Samson's improving. . . ." She touched the top of his head; the bump was smaller this time.

Although he hadn't paid too much attention to the man with the motion-picture camera, the event had been filmed and spliced to the footage of his duplicating Kronas's act. The whole thing was cut together as a news short and shown along with the silent movies at the Quinn Theatre on Houston's Main Street.

That same week a promoter named Louis Nussbaum came to town to ballyhoo a film entitled *Rosie of the Tenements*. He went to the office of young attorney Harry Dow to transact some business, and parenthetically mentioned that along with his film at the Quinn he had seen the damnedest thing—a man pulling three cars with his hair.

"I saw it, but I still don't believe it," said Nussbaum. "It's probably a phony."

"No, it's not. He's the genuine article," Dow replied.

"How do you know?"

"He's my client." The lawyer was Joe's counsel on contracts and matters relating to his garage business. "He's a wonderful guy . . . but he's a little scary."

"How do you mean . . . scary?"

"You'll see." Dow chuckled and wrote down the address. "Go look him up."

A Kissel speedster with Nussbaum at the wheel rumbled into the station. Nussbaum, a W. C. Fields look-alike, got out and, brushing a speck of road dust from his white Panama suit, looked around the homey but dilapidated station as Joe ambled over in mechanics' coveralls.

"This where I find Joe Greenstein?" he asked.

"That's what the sign says."

"Is it true what they say about him?"

"What do they say?"

"That he's so strong."

"Yeah, well . . . he's pretty strong."

"Where'll I find him?"

"Right here."

Nussbaum looked around. "Where?"

"You're talking to him."

Nussbaum had been awed by a tall character on the moving-picture screen, and because Joe's long hair was as usual pinned up in a little bun in the back, he in no way looked like the man Nussbaum had seen.

"A trifle small, aintcha?" He could hardly take the little grease monkey seriously. "Look, kid, I got no time for games here. I'm lookin' for the Greenstein who tears horseshoes up, stops bullets with his head, bites through nails, and pulls cars down the street by his hair. . . ."

Joe terminated the conversation; he went over to a Model T that had been jacked up with a flat tire, and demonstrated his latest feat—Rriipp!!!, he tore the tire from the rim with his bare hands. The promoter tipped his straw hat back on his head and scrutinized the garageman more carefully this time. "Geez, is that on the level?"

"Naw," said Joe, "I do it with mirrors."

He went about his business as Nussbaum began to tag after him.

"Listen, what else can you do?" Joe mentioned a few of his feats. Nussbaum fell silent for a moment, then reached into his breast pocket and produced a business card. "Mr. Greenstein, how'd you like to be in vaudeville?"

Joe Greenstein was a trusting soul, but he wasn't stupid. While this man's proposition sounded sincere, the sizable earnings mentioned were only half believable; even as he accepted the offer and shook on it, he hardly considered the arrangement one that would materialize.

Nussbaum said he would make contact again in six months, after he had ended his promotional tour in California. He promised to cable instructions and money for a first-class ticket to New York, and suggested that Kid Greenstein use the time to develop an act.

Having been thus "discovered," Joe returned to the garage. In the beginning, he had never considered the professional possibilities. These feats were a preoccupation, not a livelihood. He had been traveling this road all along, he concluded. He had been developing as a professional strongman his entire life, and hadn't known it. From the first moment he had met Volanko, his training, his way of thinking and living—all had been directing him toward this one pursuit. He could perform many unusual feats of strength, but now he would perfect and organize them. He would develop his skills and abilities into an act. Not sure that he would ever see Nussbaum again, he nevertheless singlemindedly threw himself into the task.

THE MIGHTY ATOM

The Golden Age of Strongmen had captured the imagination of the world for twenty years between 1890 and 1910, when the common man made his bread by the sinew in his arm and the brawn of his back. Working families were an enthusiastic and ready market for the handful of exhibitors who could perform flamboyant feats that were, in the eyes of this audience, seemingly beyond doing.

Later, in America and on the Continent, the professional strongman would work circuses and fairs, carnivals, sideshows, music halls, and vaudeville, keeping alive the sense of wonder, of man triumphant over ponderous weight and the resistance of steel bars, seemingly unfettered by the laws of physics. Into the 1920s, the strongman continued as a living wonder of the world, an awesome and inspiring vision that could be had for a modest price of admission.

Warren Lincoln Travis of Brooklyn, New York, could lift a platform holding twenty-five men in overcoats with his back, a weight of 4,200 pounds. He could lift 667 pounds . . . with one finger.

Louis Cyr, the Canadian Samson, would hold back two pairs of harnessed horses attempting to move in opposite directions. He could push a loaded freight car up a grade. He would lift to his shoulder a barrel filled with sand and water weighing 445 pounds . . . and do it with one hand.

The American, G. W. Rolandow, would put three decks of playing cards together in one stack, then tear them apart with his bare hands. A virtuoso of spectacular feats of power and agility, Rolandow would, while holding a barbell of 200 pounds, jump back and forth over the bar.

Louis Uni, a Frenchman known professionally as "Apollon," performed a stage act in which, in lieu of a conventional barbell, he employed a pair of railway-car wheels on a thick axle, an aggregate weight of 366 pounds.

Emil Naucke, a German, could hold an automobile over his head.

John Grunn Marx, "The Luxembourg Hercules," would take several large horseshoes and in quick order break them in half with his bare hands.

Arthur Saxon, a German known as "The Iron Master," could raise a 318-pound weight overhead . . . with one hand. He would lie on his back and, with feet in the air, balance a plank with twenty-two men upon it.

But it was Eugen Sandow, an East Prussian later to become a British subject, who would raise the strongman's craft from one of brute strength to the highest level of art and showmanship. It was said that Sandow was first discovered by the Vienna police when they were asked to maintain surveillance of the cafés because an unknown culprit was breaking the "test your strength" machines that were popular there as amusement. A young and well-dressed man whose given name was Frederick Mueller was observed as he deposited his coins and squeezed the handles until a succession of machines exploded. He was arrested for vandalism, but later released when he explained that the sign had said "test your strength." He had deposited his money and done so. There had been no warning that he was to stop. After apologizing for the damage, he politely suggested that perhaps the machines should be built of better materials. Frederick Mueller would become Eugen Sandow.

He was not a large man at 5 feet 7½ inches tall and 186 pounds, but was possessed of a classically beautiful physique and the catlike

ABOVE: Louis Cyr, the Canadian Samson. Courtesy *Strength and Health Magazine*

RIGHT: "The Saxons, a Trio of Muscular Marvels": Arthur on far right. Courtesy *Strength and Health Magazine*

Louis "Apollon" Uni. Courtesy *Strength and Health Magazine*

The Great Sandow. Courtesy *Strength and Health Magazine*

grace of a master gymnast. Sandow would perform a back somersault over a chair while holding a thirty-five pound dumbbell in each hand. He would perform a one-arm chin with a grip by any single finger of either hand—including just the thumbs. He would walk across the stage with a pony weighing 350 pounds upraised on his right arm. He would hang upside down from a column, then bring himself to an upright position while holding another man. He would make a human bridge of himself by supporting a plank from chest to knees and allow a horse and rider to walk across it.

The diversity of the feats executed by the Great Sandow proved him to be much more than a man of strength, and performing theatrically from 1890 to 1905, he eclipsed the others of his day.

The feats of Sandow and Cyr, Apollon, Saxon, and Rolandow had aroused incredulity when performed and would grow with the years and the telling. It was this competition, comparison with fanciful memory and ineffaceable adversaries, that Joe Greenstein accepted when he embarked upon a career as a professional strongman. With the exception of Sandow, this had been a trade dominated by giants of various configurations. Louis Cyr, for instance, while only 5 feet 8½ inches tall, had performed at a body weight of 315 pounds. Apollon weighed over 300, and Emil Naucke's last exhibitions had been at a body weight of 500 pounds. There had never been a miniature strongman of any popularity because he would, almost by definition, be excluded from competition with these giants. There were notable small men in the world of strength, individuals of up to five foot five in stature and 145 pounds in weight, among them Abe Bosches, Oscar Matthes, Otto Arco, and Maxick—Arco and Maxick being so notable because they were among the first three men in history to jerk double bodyweight overhead. Still, these muscular marvels were forced to pursue careers in allied fields in which their diminutive size would not be a limiting factor. Some were wrestlers, others posed as artists' models (Arco modeling for sculptor Auguste Rodin) or performed gymnastic or strength acts of other kinds. Arco became one of the finest balancers in vaudeville. However, these small men did not become strongmen, where at feats of lifting, bending, and breaking they would be inviting self-defeating competition from men more than twice their own size. It was typical of Joe

Greenstein that if any of these barriers concerned him, he did not allow them to alter his course.

In 1927, the arena to which the American strongman aspired was that most popular form of entertainment, live variety theater—vaudeville. There, unlike the tawdry carnivals and sideshows to which strongmen were increasingly relegated, exhibited between a tattooed lady and a set of Siamese twins, he could perform in a legitimate show with his act a part of an evening's wholesome family entertainment.

Joe carefully reviewed his training to ascertain how it might be used on the vaudeville stage. He took to breaking chains with his

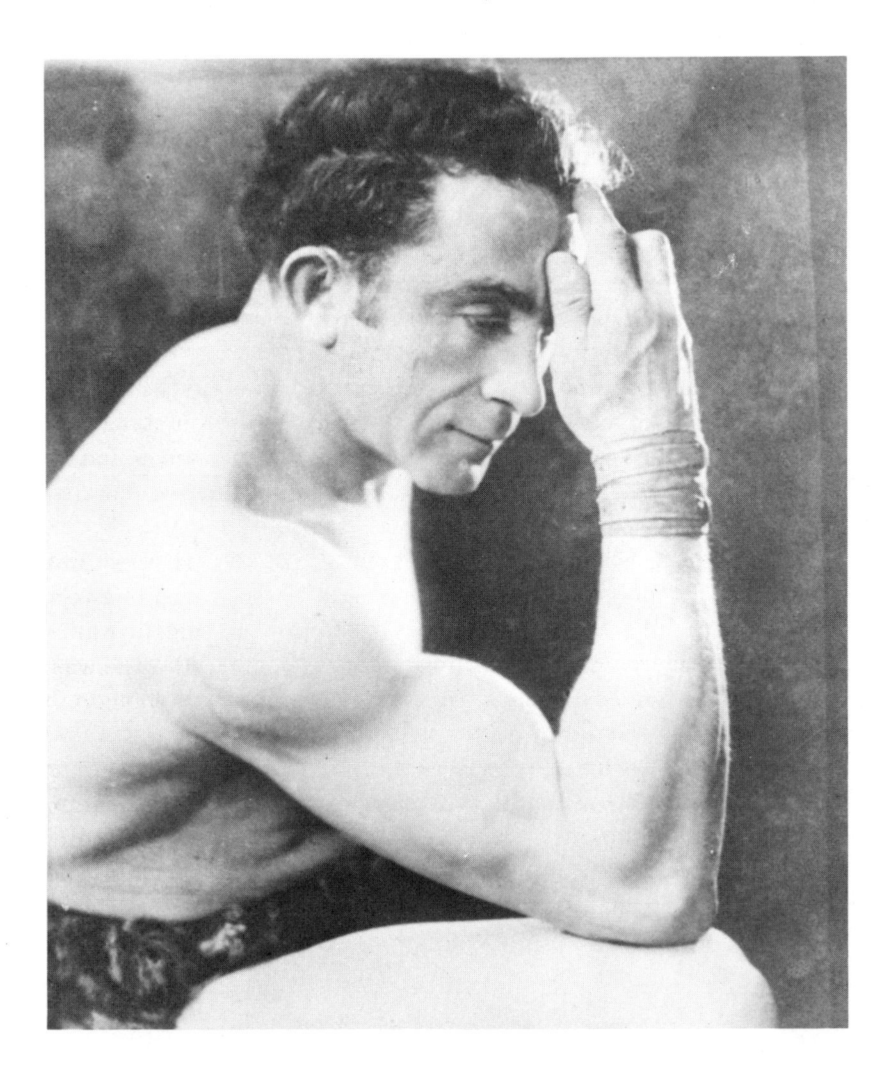

chest, having advanced Volanko's lessons of years ago, and perfected both high and low breathing. Now, taking a low abdominal breath, he would hurriedly put a protective heavy leather belt around his deflated chest and fasten it tightly with three lengths of number 8 s-link jackchain. With a supreme effort, he would rapidly transfer the air from stomach to chest, his 13½-inch chest expansion exploding the chains.

He had toughened the neck muscles and tendons to where now a bar of cold rolled steel 5 feet long and 1 inch in diameter could be bent across his throat without ill effect. He could do the same over the bridge of his nose, with the aid of a small formfit aluminum nosepiece notched to hold the bar.

Biting techniques had been practiced professionally in modern times by only one other strongman, the great Siegmund Breitbart. But not even Breitbart could master such adamantine objects with this specialized feat of strength; Greenstein could bite dimes and quarters in half, and with the aid of his thumb applying pressure from the side, could even crush silver half dollars with his teeth. He began working on other even more bizarre stunts of jaw and neck power.

His Samson routine had developed to where he could tow as many as five cars at once with his hair. The bump on his head had long since stopped showing up. Knowing that Breitbart, Kronas, and others had used the bed of nails, he too constructed one. He theorized that these feats were performed by muscle contraction and proper use of leverage, and he devised and executed numbers similar to what the others had done.

While other vaudeville strongmen might have flashy costumes and trick showmanship, he wanted none of it. He wished to awe an audience with sheer dynamic power. And one more thing: he wanted no one to doubt his honesty. Therefore, he vowed as long as he was a performer to make a practice of having anyone who thought he could duplicate a feat come up on stage and try it.

Greenstein did not disparage the lifting of weights, but would not use them in his act because they were not his style. Some strongmen do not lift weights, some weight lifters are not strongmen. Bending, breaking, biting, driving, twisting, Joe's strongman's stock-in-trade required different techniques, and sometimes even different muscular development. Kid Greenstein was not a weight lifter. He had trained with them to be sure, though he rarely employed more than twenty-five pounds with either hand, preferring the results of constant repetition to the strain of ponderous weight. However, in the garage he did keep a barbell of 163 pounds. He could curl and press this weight over his head with his right hand—gripping the weight by only the middle finger. A callus developed on the bottom of the middle and the tops of the second and fourth fingers as he perfected this lift. A one-finger curl and press of 163 pounds is extraordinary at a bodyweight of a mere 145 pounds, but he found weights to be unimaginative and his lifting abilities too narrow to be included in an act.

There were also other feats of strength that he would not use in vaudeville, for all their seeming drama. Before a crowd at Houston's Pearson Ford plant, he had held back two automobiles. His hands grasped a single iron ring, and a rope was looped around each arm at the crook of the elbow, the cars were given the gun, powering in opposite directions with him in the middle. Their wheels spinning, they remained in place; his arms did not break their hold. Imitative of Louis Cyr, he later successfully tried the stunt with two horses. But weights and pairs of horses or cars made expensive and unwieldy props. He preferred the kind he could carry in a suitcase.

After six months had elapsed, a hoped-for but unexpected cablegram arrived: "LEAVE AT ONCE FOR NEW YORK—NUSSBAUM." With it was one hundred dollars for first-class passage and expenses.

"Leah"—he showed his wife the money—"that fellow Nussbaum just sent me a hundred dollars. He wants me to come to New York and go on the stage as a strongman."

"Is that what you want to do, Joe?"

"Mom, for the longest time I've been onto something very important. I've got this gift, but I've never quite understood what I'm to do with it. I think this ability I have is wasted unless I can show it to people. If I'm going to demonstrate seriously, I'll have to go to New York."

"Then do that," she suggested.

"But how can I leave you here with all of this?"

"Don't flatter yourself, wise guy," she said. "The kids will pitch in pumping gas and helping out around the house, and I'll take care of the rest. If you make a go of it in New York, we'll come to you. If not, you'll come home, and we're no worse off than we were before."

"Leah, being apart from you and the kids . . . it would be terrible."

"Yosselle, you work fourteen hours a day as a mechanic, and if your hands bleed you don't complain. So now if your heart tells you to take one more little chance, well . . . maybe it's better than to wonder what might have been."

"Nussbaum is a promoter and I don't know if I can trust his word, but people make big money in vaudeville, and I guess there's only one way to find out."

"If it's what you want, Pop," she said, "go and try. I'll pack your things."

The next day all nine children were washed and assembled in the living-room gym. He kissed and hugged each one. "Judy . . . Harry . . . Esther . . . Isaac . . . Minnie . . . Mary . . . Moishe . . . Mendel . . . Rivie . . . I'm counting on you. While I'm away you take care of your mom, and look after each other, okay?"

He held Leah at the door.

"Don't make a big deal. We'll be fine."

"I'll send money. If there's any problems, wire me right away, and I'll come home."

"Joe," she said as an afterthought, "don't take any chances you don't have to."

Having said his good-byes, he caught the train for New Orleans, and from there a boat to New York. He boarded the vessel carrying luggage containing a bed of nails, several dozen assorted horseshoes, bars of steel, railroad spikes, and a genuine leopard-skin strongman's tunic which had been made for him in a tailor shop on Houston's Main Street.

He felt his act needed polishing, that he wasn't ready, so with the permission of the ship's captain, he decided to present a free show for the passengers. If he was going to flop, he might as well do it anonymously. To his amazement, after a few iron-bar and horseshoe-bending routines, the people were throwing not tomatoes, but money. Still, he had to remain cautious; it might be beginner's luck.

He arrived in New York on a mild evening in the spring of 1927, and after checking in at a cheap midtown hotel, took a walking tour of the Great White Way. He found himself in the theater district: block after block of rococo palaces with pilastered lobbies and lighted marquees were butted one against the other. These were not vaude houses but legitimate theater, and he wondered how these stage plays had been afforded "legitimate" status with what seemed to him prurient subject matter:

"SEX" with Mae West at Daly's Theatre
"SCANDALS" at George White's 42nd Street Apollo
"SINNER" at the KLAW West 45th Street
"PUPPETS OF PASSION" at Chanin's 45th Street
"SCARLET LILY" at the Comedy 41st Street, and
"A VERY WISE VIRGIN" coming soon to the Bijou on 45th Street

Joe doubted that she would be a virgin when she arrived, what with advertised carnality extending even to Broadway billboards where the Hudson Motorcar Company extolled "The Thrill of the Super-Six . . . Freed to the Limit," and a mere cigar like Robert Burns promised ". . . Complete Satisfaction."

Despite his travels, Joe Greenstein had been small-town all his life. Now, he stood at Broadway and Forty-fourth streets captivated by the polyglot stream of life that passed by him, and he concluded that New York was not so much different as it was more, every kind of human being and endeavor in amazing multiplicity. If he was going to succeed as a performing strongman, this was the one place to do it. He had found the center of the world.

On the way back to his hotel, he paused in front of the Roxy, "The Cathedral of the Motion Picture," the world's largest film theater which had just opened at Seventh Avenue and Fiftieth Street, and he stood there thinking of his Houston Sundays. On that day the children would excitedly line up two abreast in size places, and with Leah leading the troupe carrying a baby in her arms and he behind with his Samson's hair flowing, the family would march down the street to Thompson's Café, and from there on to the movies. No matter what glories, if any, there were to be garnered in New York, he would miss those Sundays most of all. But he vowed that this time would be well spent, that his succession of disappointments would end here.

The next morning he called Nussbaum, and at his instructions went to 727 Seventh Avenue, to the office of Ab-Gold Film Corp., a firm with which the promoter had an association. The name Ab-Gold was a contraction of the names of the partners—Abramson and Goldberg.

He met Nussbaum at the Ab-Gold office and together they walked into a waiting room filled with various kinds of talent: a brace of ham actors, a coven of magicians, some cowboys, a pirate, everything it seemed but a gypsy and a monkey. A pushy sort, Nussbaum stormed past the secretary into the inner office where Ivan Abramson, a lanky, bespectacled man of studious appearance, and Goldberg, a tree stump in gartered shirt sleeves, were doing business.

"Split weeks, my ass!" Goldberg bellowed into the phone. "The Sparrow Lady don't play no friggin' split weeks!" He slammed the phone down.

"Louie . . ." Abramson welcomed them.

"Joe Greenstein," Nussbaum introduced Joe, who looked like a bumpkin in his ill-fitting suit.

"You didn't say what he does, Louie," Abramson inquired. "What's the secret?"

"Whaddya do, kid?" Goldberg asked. "Sing? Dance? Juggle?"

"No, nothing like that," Joe explained. "I do things no one has seen before. You see, there's a higher power. Everyone has it, but I'm learning to harness it."

"What do you *do?*" Abramson asked pointedly.

"I guess you'd call them feats of strength," Joe answered.

Goldberg looked him up and down, "A strongman?"

"Well, yes," Joe clarified, "but the power is more mind than muscle. . . ."

"Mind, schmind," Goldberg interrupted. "Forget it, Louie. He's too small." He turned to Joe, "Whaddya gonna lift, kid . . . peanuts?"

"No, I don't think so," Joe said; he sat a case of paraphernalia on their floor and went to work. Twenty minutes later, as Joe stood amidst the mangled iron remnants of his performance, Abramson and Goldberg looked at each other with disbelief. The office came alive with activity; Ab-Gold Film Corp. knew a good thing when they saw it. This little immigrant had tremendous potential for the vaudeville stage.

"THE PREMIERE APPEARANCE OF JOSEPH L. GREENSTEIN."

The name would never do. "We need a vaudeville handle," Goldberg insisted. "Something with flair . . . style," said Abramson. The office activity ceased as quickly as it had begun, and all sat down to conjure up a new name. The others bandied suggestions about as Ivan Abramson quietly studied Greenstein, then spoke up after a long thoughtful moment. "The Great Atom," he said.

It was perfect: appropriate, oddly descriptive, and with an imposing ring to it. After a few weeks and some reflection, Abramson would suggest a refinement. "Great" was cliché, but "mighty" was eternal. "The Mighty Atom."

It was some weeks before Abramson, Goldberg, and Nussbaum were satisfied that Joe was ready; but now they had no intention of booking him on the vaudeville circuit without a proper introduction. With an act like his, they wanted to make a big splash for openers. Sponsored by the New York *Journal* with all proceeds to the Cynthia Grey Children's Milk Fund, the newly dubbed Mighty Atom found himself dressed in his strongman's costume at Broadway and Forty-sixth Street. It was June, 1927.

He stood outdoors, behind a makeshift rope barricade, while cordons of police controlled crowds and traffic, and dignitaries made their speeches for the milk fund. The Mighty Atom prepared to make his New York debut by pulling three heavy trucks full of children, a combined weight of over twenty tons, down the Great White Way, with the lead truck tied to his hair.

Times Square settled into an eerie quiet as he began. Then, a roar went up as he got the three vehicles moving; but, several yards into the pull, the enormous strain of the tonnage broke the comb and locks, and the cable sprang away, slithering across the street, its end shooting sparks as it came to rest at the trolley tracks.

Disconcerted, Greenstein's one thought was to complete his task. He ran over and knelt to pick up the cable; someone whacked him on the arm with a billy club. The strongman turned angrily, but checked himself at the sight of a burly uniformed police sergeant. One from the crowd called out, "He just saved your life, Mr. Atom." In those days, New York's electrified trolleys were powered from the ground up through a narrow slot by a third rail under the street. Joe knew nothing about them. The end of the cable lay inside on the third rail; to have picked it up would have been instant death. With appropriate thanks to the policeman, and the cable retrieved and safely retied to his hair, the Mighty Atom completed his march down Broadway.

A large tent had been erected on the cordoned street and after his performance the Atom was installed in it to receive the public. The multitudes jammed through the turnstiles to meet the modern Samson and dropped their contributions into the milk-fund collection boxes. Each man, woman, or child who came to see him predictably had the same idea; each shook his hand and squeezed to test his strength. After exchanging greetings with several thousand New Yorkers, by the end of the day his hand was so traumatized that he couldn't lift it.

In Joe's mind, performing for the milk fund had been like leaving the door open to Your Filling Station; he expected no reward. He did not know that Ab-Gold and Nussbaum had prepared to get publicity from the event, so it came as a surprise to Joe when he found himself lauded at City Hall, and like other vaude performers and dignitaries, presented with the keys to the city by Mayor Jimmy Walker. In another town such an honor might have been the key to easy street; but in New York, news a day old is no news. A week after receiving the keys to the city, Joe found himself in relative anonymity once again. To him it was extraordinary that fame disappeared as instantly as it was gained. Show business was now even more of a mystery to him.

He began rehearsals at the Mount Morris Theatre on 116th Street and Fifth Avenue near Mount Morris Park. For four hours a day, onstage, he developed his act under the supervision of Ab-Gold and Nussbaum. In his hotel room at night he would labor another four hours. He had already perfected his old numbers and was in the process of perfecting new ones for inclusion in the act. Though he missed Leah and the kids, he forced himself to think only of the job at hand. He did little else as he attempted to exceed his known limits with more and more difficult feats of strength.

The Hebrew mystical writings, the Kabbalah, refers to "the Inner Light." The Orientals have a name for inner strength: "chi" in the Chinese, and "ki" in the Japanese. It was the quest for this power which filled Joe Greenstein's hours and days. This spirituality which he sought was pragmatic, for the higher his understanding, the greater the result in his physical performance. His muscles, bones, and tendons had been tenaciously prepared, but they were only the vehicle. It was this life-force which drove the machine; he had trained the machine not to break.

He had attempted to scientifically analyze his strongman's feats, but science failed to give him answers on several levels. Seeking to chart a course, he continued with his studies of Jewish mystical writings, fascinated in particular with the Kabbalah. According to the Jewish tradition, no scholar may study Kabbalah before the age of thirty, because it is said that one who immersed himself prematurely in conceptions of matter and time which were beyond man's grasp, and the attendant numerology by which these were interpreted, could be in danger of losing his mind. According to the Zohar, one of the two central classic works of the Kabbalah, man is a microcosm embodying all of the mysteries of the universe. Joe knew this to be true, though the writings indicated that the Infinite Light was beyond human understanding; still, he sought answers. Revelations had been given to Jews throughout history, and now this Jew searched for the revelation within himself.

> . . . FOR EVERYTHING WAS FILLED WITH
> THAT SIMPLE, BOUNDLESS LIGHT,
> AND THERE WAS NO SUCH PART AS HEAD,
> AND NO SUCH PART AS TAIL;
> THAT IS THERE WAS NEITHER
> BEGINNING NOR END,

FOR EVERYTHING WAS SIMPLE OR SMOOTHLY
BALANCED EVENLY AND EQUALLY
IN ONE LIKENESS OR AFFINITY,
AND THAT IS CALLED THE ENDLESS LIGHT.

The Kabbalah
The Tree of Life
Contraction and Line of Light

The harnessing of the Light would not be a demonstration of a destructive force, as the splitting of a chain or the crushing of a steel bar might appear; but a materialization of that energy which, while at the very source of the universe, was unbridled in the human being. Joe's concepts were esoteric to be sure, but his life to date had been an amazing rite of passage. He had reached the juncture where, with concentration, he could summon a measure of the force at will. He was amused by those to whom he disclosed a small part of his introspections who looked at him askance, or derided these as fanciful notions. This Light was not Joe Greenstein's; it emanated from within every human being, and animals, and plants, it was in the sun and the air, it filled the farthest reaches of the cosmos.

The application of this force to feats of strength was something that no one could teach him, but Joe knew it could be learned. The power had to be compacted for it to be applied against objects. Unlike a wave of water whose broad brute strength crushes everything before it, Joe strove to employ himself as a conduit, to focus this power into a minute pinpoint of irresistible energy . . . a kind of human laser beam. The beam would not assault the entire object, but concentrate upon its center, its core. The rest would follow.

For him, the bending of metal became a spiritual endeavor. He would invest the inanimate object with a being, a character of its own, as if it were his adversary. "I am Man," he would say, peering at it. "I am possessed of the Power. You are metal . . . without will. My will is superior to you. The Power will overcome you. You will bend . . . you will break. . . ." An indescribable impulse, a wave of energy, would sweep over him, as if he were no longer himself but something much greater. He could feel it being transmitted out of his eyes and converging into the shiny steel, feel the waves of it over his face, coursing through his hands. And at the zenith of this moment, when he had pitted his very being against that centerpoint of the object: ". . . you will give way . . . NOW!"

The mind commanded, the body reacted, and the object inevitably succumbed. It was just as Volanko had taught him: once an action began it had to be irrevocable, beginning and end in one thought and motion, like a bullet fired from a gun. Experience dictated that the difficult steel bars had to be mastered quickly . . . or not at all. Speed insured success, while delay threatened failure. He observed that the first bend actually weakened the bar, but once that was accomplished the rest had to be done immediately, or the steel would become even stronger than before. He theorized that the heat and friction of the bend in the bar repelled the resistance to its ends; the crushed bars were hot when he tossed them aside. Every maneuver was practiced again and again, bending and breaking the bars, increasing the gauge as his technique and concentration improved, devising the new and unusual, putting the stamp of the Mighty Atom on feats of mind over matter. As a good blacksmith communes with his materials, so did Joe develop the ability to merely look at a piece of steel, to hold it in his hands, and know the energy necessary to master it.

He felt inwardly that he had been fated, chosen to do these things. "Gibor Y'Israel." Strongman of Israel. That Inner Light mentioned in the Kabbalah was a vision in his mind, ethereal but very real. The Mighty Atom would not show himself as a creature of dumb strength, but as one who sought a higher spiritual and physical plane. He would do this humbly, as much by fasting as by eating, more with brain than with biceps. The whole would be more than the sum of its parts. By performing these superhuman deeds he would prove that this life-force existed, that there is no such thing as a little man, and nothing is impossible.

The Mighty Atom made his stage debut at the Mount Morris Theatre on September 9, 1927.

"Leah," he wired home to Texas, "I'm in vaudeville. They like me. . . . Love Pop."

It was small-time vaude, but by no means an inauspicious beginning. Ab-Gold had gotten the push they needed and booked him nonstop for the next several months.

The money he sent home was more than he had made as a simple working man, but the sums he received were small when compared to the figures that he presumed he was earning. He questioned Nussbaum about the net seeming to be only a small percentage of the gross. "Expenses," Nussbaum explained, and after a quick rundown of the numbers involved, Joe was more confused than when he began. He accepted the checks they gave him.

Living in a dollar-a-day hotel room with faded flower-print wallpaper, Joe knew a part of his life had drawn to a close, and another had just begun; he would never return to Texas. With feelings of uncertainty mixed with hopeful expectation, he sent his next telegram:

"Dearest Leah, kiss the kids for me. I am working. Close the station and come to New York. . . . All my love, Pop."

As he waited for his family to arrive, the Mighty Atom labored to perfect two new and remarkable feats for exhibition, "The Questionmark" and "The Corkscrew." One looked impossible, the other rather easy, but actually they were equal in difficulty.

For the Questionmark, he would take an enormous regulation No. 5 horseshoe and fasten one end of it at shoulder level into a standing vise which had been brought on stage. He wrapped the other end of the horseshoe in a soft handkerchief. Gripping the horseshoe in his front teeth, he would push up with neck and legs, twisting the iron. Changing the horseshoe's angle in the vise, he would bend it again, contorting it into the shape of a questionmark.

With his hands, he could bend a No. 5 into a "W" or break it altogether in about a minute. The harder a horseshoe is, the easier it is to break. A softer one may have to be bent back and forth several times before it will snap.

It was a matter of personal honor that his act be an honest one. He would never gimmick the temper of a horseshoe or anything else used in his act. The very idea of rigging a prop was unthinkable. He bought regulation horseshoes from hardware stores; a barrel of about a hundred cost a few dollars.

Greenstein had acquired immense strength of arms and hands by his constant practice of bending and breaking iron. Now, he ordered ½-inch-square cold-roll steel bars and practiced bending several a day, using shorter bars each day.

The Greek mathematician Archimedes had said, "Give me a lever long enough, and a fulcrum strong enough, and single-handed I can move the world." The converse is also true. With a lever too short, you can do nothing. It was this lack of leverage that made the feat extraordinarily difficult, and it was this that Joe finally mastered. Delivery men brought boxes of short steel bars to his hotel room, taking away bars from a week before, now mangled into pretzel shapes. Each delivery became a little lighter as the bars were reduced in length. He could now crush ½-inch thick bars of only 9 inches in length.

With this process of bending short bars, he developed a pronounced muscle on each hand at the base between the thumb and forefinger. What was on most men a small patch of flaccid flesh, he could flex as hard as a walnut.

The physical and mental power that he had gained allowed him to attempt "The Corkscrew." This, of all the hand-bending and -breaking routines in his regular act, was by far the most difficult. To the layman, it looked rather simple, but it was for him a supreme test of his resources.

One end of a flat steel bar 1¼ inches wide, ½ inch thick, and 16 inches long was fastened in the standing vise. With only a thin handkerchief as a cushion, he grasped the free end of the steel with both hands and turned. The effort necessary to make the three twists of the bar was a terrible test of resolve. "My insides turned over as much as the iron," he said about it. After three twists, the flat steel resembled a corkscrew.

In all the times that he would perform this onstage, he would never even receive a round of applause for it. It simply looked too easy. He would get standing ovations for much lesser feats; but for this one of gut-searing difficulty . . . nothing. It reaffirmed his belief

that most pinnacle accomplishments are lost on the average man.

A number which audiences appreciated more, but actually required less hand and arm strength, was his "clincher tire" routine. He had first performed it in his Texas filling station, but perfected it in New York. A protruding bead in the edge of the tire fit into a groove in the clincher rim, making separation quite difficult. A large passenger car was brought onstage. After lifting it up without benefit of a jack and placing its front end on a milk box, he would demount the clincher rim and deflated tire as a single unit. Then, with bare hands, he would rip the tire from the rim, and replace it, all without tools.

He began integrating "decorative iron" routines into the act. The same 1¼-inch-by-½-inch steel that he used for The Corkscrew he employed in an 8-foot length, wrapping it around an arm or leg until it resembled a coiled spring.

The Mighty Atom accepted the premise that there is no such thing as a bad audience. He was doing his turn for a particular matinee crowd, and music briefly played him off before his return for the finale, but he was frustrated that the audience had remained largely unmoved.

"Is this a home for the blind? What do they want?" he said as he reached the wings.

"Christians and lions," wisecracked a knowing stagehand.

Joe had one number in his repertoire that he had performed onstage, but only on those occasions when the audience was not reacting up to his standards. He would do all of his feats before attempting this one, and then only grudgingly, as a last resort. In difficulty, it surpassed the Corkscrew, and it would deplete him for the rest of the day.

Music played him back on, and he signaled the substitution to his MC as a soft drum roll began. The Atom summoned his inner power for what was coming. Unlike other feats, this one never got any easier; it would wring the last ounce of resolution from him every time.

He held up a long length of welded-steel tow chain, then swung it around his back and fastened it to his arms. The sonorous hum of the snare drum built slowly, Joe's face set with total concentration as he tortuously flexed his arms forward, pitting himself against the steel chain. After a minute, the drum roll swelled to a staccato thunder and at its height a sound was heard like the crack of a bullwhip splitting the air; the chain burst, its ends flying out, the force of Joe's momentum carrying him forward pitching him off the stage into the orchestra. The audience came to its feet with an ovation, and Joe scrambled out of the pit to take his bows.

There would usually be one Samson stunt per show, such as pulling a ton of grand piano with a dozen men atop it across the stage or swinging a "merry-go-round" of four women by his hair.

To dispel any doubts about the authenticity of the Atom's steel objects, ten men from the audience were brought on stage and seated. The bars and horseshoes were passed down the line, each man examining or making an attempt at bending it. With one assistant presenting the items and the other at the end of the line collecting them, the Mighty Atom received them at short intervals. As the band played, and the committee authenticated a new object, the previous one had already been quickly bitten, bent, driven, twisted, or crushed.

The average turn for a vaudeville performer was about ten minutes. "The Mighty Atom" would be the climactic, closing act, and it might last as long as eighteen minutes. There would be particular interest in his finale—people don't walk out until they get the best of what they came to see. A ten-man Dixieland band would be put on a ton-and-a-half pick-up truck. The vehicle would be pushed up onto a wooden "bridge" which had been placed atop the Mighty Atom's chest . . . all while he reclined on the bed of nails. At the apex of the bridge the truck would stop for a half minute while the band played "Stars and Stripes Forever." At his signal, the truck would then roll off slowly.

Greenstein explained this feat in the most logical of terms: "If you take a sharp knife and tap it on your arm, or even press a little, it won't cut. Pull it . . . it cuts." Therefore, the idea while on the bed of nails was to take the pressure without any unnecessary or sudden movement. Everything slow and easy.

It was critical that the driver of the truck do all things smoothly.

He was to apply the brakes only once at the apex, and not to jerk the machine or relax his vigil until the feat had been completed. The machine would have to be pushed manually and could not be driven under its own power, since gasoline was forbidden inside the theater. Joe would talk the driver up onto the bridge—"Easy . . . easy . . ."

The bed of nails in itself was not the greatest danger. The more nails there are in the bed, the more support and the safer it is. The most danger was in the weight. The mystery of a man taking several tons on his chest can be explained by his technique of breath control and use of leverage. Joe would take in air, hold his breath, and contract the muscles, remaining still and in control until the pressure from atop had passed. He would have to hold that breath with the weight on him, knowing that if for any reason he exhaled, he would be squashed and the nails would puncture him. Other performers had used "The Bed of Nails," but few with the degree of risk that the Atom attempted. If no truck was available, he would stand twenty adults on the bridge atop his chest. The six-piece metal and wooden structure itself weighed 450 pounds.

Scrupulous about safety precautions, his great worry was blood poisoning. Even a tiny pinprick from a single nail might be fatal. Such, it was reported, was the fate of the great Jewish strongman Siegmund Breitbart. Among other of his feats, Breitbart was the only other man in the world who could bite through a steel chain with his natural teeth. This young man of superior power, the most popular strongman of his day, received a slight scratch on his back while performing on the bed of nails. No man was stronger than the lockjaw germ, and he died shortly thereafter.

Fearing a similar fate, Joe Greenstein carried copious amounts of peroxide with him wherever he traveled; before and after every performance, he scrubbed his back with the solution to ward off infection. Exhibiting "The Truck on the Bed of Nails" two or three shows a day, he observed all the precautions and escaped without injury.

For all of Joe Greenstein's incorporeal means of overcoming insuperable objects, there were less lofty methods which he sometimes employed at the same time. A preconditioning process would begin in his dressing room where he would sit and scrutinize a short, wide bar of steel or a horseshoe more challenging than the most difficult he had mastered. He would then air any unconscious excuses or evasions, honestly appraising any reasons for failure. "Do I have any physical problems which will prevent me from achieving this?" he considered. "No." He went on. "Any psychological weakness? Family problems? Lack of resolve or any concern which I might allow to interfere?" More serious consideration. "No. Problems of leverage, arm length? No." He looked at the bar. "It can be done. But must I really do this? Is it, at this moment, more important than my own life? *Yes*."

Once he had examined all the reasons for failure and found none that were substantive, and had infused himself with the will to persevere, he went to the next step, putting himself in a position where there was no alternative, no way out. He announced to his audience that the herculean bar was his next feat. Failure was now out of the question; he had burned his bridges. There was only one way . . . forward. The bar would have to go . . . it did.

There were, inescapably, times onstage when even the Mighty Atom was exhausted, when his personal resources faded: and at these times he would sensitize himself to the presence of the audience. Their eyes were on him, he could feel the energy of a thousand eyes beaming at him in unison, and he told himself that he could draw the power from them and redirect it to the object at hand. He also used this mental set at times when his act was going very well and he wished, without the usual preconditioning, to master steel that had heretofore been beyond him. In this manner, he found himself bending with ease objects that he had not been able to budge an hour before in his dressing room. Whether imagined, actual, self-hypnosis, or psychic gamesmanship, Joe felt in these moments that he and the audience had overcome the object together, but he never told them.

Despite the seemingly superhuman feats of strength that he was able to perform, there was something austere about the Atom's act. It was impressive, but lacked razzle-dazzle. As a strongman he was a professional, as a showman he was a steadfast amateur. When others suggested glittering gimmicks to please the balcony, he dismissed such as nonsense. He was told that people would be entertained more by phony hokum than the true article, but he persisted. He tried to explain to Nussbaum his higher goals, but Louie was less than receptive. "Hey Mister Goody-Goody . . . who're you? The angel of Broadway? You wanna educate? Forget it. They wanna be entertained."

The Atom's act involved little more than a clinking collection of outrageous iron and a nail barrel crudely painted silver. He was there to perform genuine specialized feats of mind and muscle over matter and to him everything else was cotton candy. Strongmanship more than showmanship.

For his part, Nussbaum was pure showbiz. Joe looked with skepticism upon some of his concepts for the act. The Atom had agreed to having two men and four shapely women onstage to assist him and help beautify the act. He owed his management fifteen hundred dollars for the cyclorama with which he traveled. But that was the extent of his compromise.

The Atom had been in New York and rehearsing for about six weeks when Nussbaum came running breathless into the Ab-Gold office. A terrific idea had come to him in the night: they would construct a dancing platform to be supported by the standing strongman and while music played, four teams of marathon dancers would dance on his shoulders at once. The Atom told Nussbaum that he didn't care much for the idea, but Big Louie was not to be dissuaded. "The platform's no problem, but the harness to hold it has to be custom made. . . ." Immediately, Nussbaum packed him into a cab, and they sped downtown to a saddle maker on Twenty-fourth Street who crafted leather rigs and harnesses for circuses and vaudeville acts. The Atom reluctantly allowed himself to be measured for the harness, but as he put his shirt back on he noticed a girl staring at him. She was an equestrian performer with Ringling Brothers.

In the early twentieth century and before, men did not train with weights to build symmetrical bodies. They trained for strength, and muscular, sometimes oddly proportioned bodies were the result.

Later, body builders would train for shape, and strength would be its by-product. Though at the time the idea of developing a beautiful symmetrical body was rather rare, the Mighty Atom had achieved just such a physique.

The girl stared at him from across the room. After a moment she motioned for him to come over. Naively, he did so. She smiled provocatively, then casually put her hand between his legs. He stood frozen, never having encountered such behavior. Mortified, he stormed out of the shop with Nussbaum's laugh braying behind him. The woman had so disconcerted him that he walked the several miles back to his hotel. For this and future conduct evading all manner of showgirls, Nussbaum nicknamed him "Angel."

Leah and the kids arrived in New York on Armistice Day, November 11, 1927. He met them at Penn Station and, with hopes that their adjustment to New York would not be too difficult, took them home to the small house he rented on Douglas Street in Brooklyn.

Brooklyn was a place in which his children would be at home in neat tree-lined streets of private and row houses, apartment buildings filled with working-class families, spacious parks, empty lots in which to create adventure. An occasional farm still dotted the extremities of Brooklyn, and it was rumored that Indians still lived in Canarsie. The typical Brooklyn corner apartment house contained a row of small stores on one side at street level; a pharmacy where the neighborhood "doctor" removed cinders from one's eye without an appointment, an Italian-owned barber shop where two minutes after a haircut, gooey green hair dressing set on a boy's head like concrete, a Chinese sanitary hand laundry, and a perennial "Pop's Candy Store and Luncheonette" where sandwiches, "egg creams" of Fox's U-Bet Chocolate Syrup, and ten-cent toys were dispensed. It was not automobiles that ruled Brooklyn streets but children, wherever one turned, boys dashing madly between parked cars shouting, "RINGALEEVEO ONE, TWO, THREE!" Brooklyn games with Brooklyn names. "Stickball," street baseball with a piece of sawed-off broomstick and a pink Spaulding ball called a "spauldeen." Local Babe Ruths claimed the distance between manhole covers . . . "*two sewers!*"

Brooklyn was a tiny nation within itself, with mini-states called Bensonhurst, Flatbush, Boro Park, East New York, Brownsville, Bay Ridge; sections divided into neighborhoods, each subdivided into myriad smaller neighborhoods, but everywhere the same, and each with its own vaudeville house. It was a place filled with immigrants—Jews and Italians, Irish and Scandinavians, all at close quarters in a borough surrounded by water and insulated from the rest of the world in the minds of its inhabitants. Manhattan could be seen from any high rooftop, but it was always called "the City," as if Brooklyn were as separate from it as England from France.

After successes at the Stadium, Stone, Supreme, Republic, Apollo, Ambassador, and other theaters of the small-time Brooklyn circuit, Joe started booking out of town. Leah accepted the situation but was concerned about his proximity to showgirls.

"Watch out for those girls, Pop." She understood the obvious temptations.

"I am impervious to the advances of any other woman," he replied earnestly.

"You'd better be." She hefted an iron frying pan and aimed it at him, only half in jest.

"My Leah"—he put his arms around her—"is all the woman I'll ever need."

She had lost her figure four children ago and her good looks in the harsh Texas years, but in his eyes she could never change. In sexual matters he clung to a simple precept: One God, One Woman. They had grown from children together, and had known no other. She was his inspiration, forgiving him time and again for his eccentricities and failures. Leah more than he was the bedrock upon which the foundation of the family was built. He could not betray the trust of such a woman, though at times the array of flawless bodies that wriggled past his view sorely tried his principles. On one occasion he turned back a woman admirer at the stage door: "I'm terribly sorry, madame. I'd really like to accept your generous offer, but you see . . . ," he said, "there's nothing I can do for you. Unfortunately, I lost it in the Great War." He made sure that Leah got wind of his nickname, "Angel," and it reaffirmed her faith. As he worked the myriad vaude houses, he tried to stay close to home, and rarely accepted engagements west of Ohio.

WRIGHT WHIRLWIND

Returning from a New England engagement to Brooklyn's Stadium Theatre, the Mighty Atom came up with the idea of creating publicity by pulling a chain of three trucks by his hair every morning in front of the theater. For the next week, the Stadium was standing room only.

Nussbaum and Ab-Gold had discussed with him the idea of including an audience-participation number. It was decided to put the Atom down on the Bed of Nails, place a 350-pound blacksmith's anvil on his chest, and have a couple of hefty volunteers from the audience come up onstage. As the music played "The Anvil Chorus," they would take turns pounding on it with heavy sledgehammers. The number went over fine until that Saturday night when one of the volunteers, from stage fright or alcoholic influence, missed the anvil and brought the sledgehammer down flat on the Atom's chest. It would have killed a lesser man, but his bones and muscles had been well-conditioned. None was shattered, but he was black and blue for months after, and had to apply makeup to his chest to cover the discoloration. Three weeks after the accident, he was still spitting up blood. The anvil number was out after that; he concluded that such volunteer numbers were more dangerous than his most arduous feats.

Playing a Long Island theater in early 1928, Greenstein experienced upsettingly sparse audiences. Poor management, second-rate acts—and the people had stopped coming. The Mighty Atom had been booked at a hefty salary, but would have to help publicize the appearance himself. The Samson stunts had aroused attention. "But, Atom," Nussbaum cautioned, "pulling cars and trucks ain't new anymore. Competition is tougher every day. Vaude strongmen are a dime a dozen. You wanna feed that family of yours? You gotta do what you said, you gotta give the people what they never seen before. . . ."

"An airplane," Joe thought out loud, "has anyone ever attempted to hold back an airplane?"

"Could you do it?" Even Nussbaum marveled at the suggestion.

"I wonder."

It would be spectacular, if it could be done. By now his hair and scalp were so conditioned and tested that his head would come off before they did.

The airplane would be the most dangerous thing he had ever tried. But if he failed or got into trouble, someone could give a quick signal to cut the engines. At worst, he might be dragged head first across the field.

As he figured it, the single-engine aircraft would have to be held from a standing start; no one could hold back an airplane once it got moving. Trucks have gears and driven wheels, airplanes don't. There would be no great traction, no powered wheels to dig in. Thus, he concluded logically, the pull should be less; he believed he could hold back an airplane shackled to his hair.

"I'm going to try it, Louie," he informed Nussbaum.

After hurried ballyhoo in preparation for the event, Nussbaum drove the Atom through Long Island farmland to Curtiss Flying Field near Garden City. They were surrounded by photographers and newsmen as the Atom made his entrance onto the runway. The aircraft chosen for the occasion was a high-wing monoplane powered by a Curtiss C–6 engine and dubbed the *Winnie Mae;* it had the same name as the aircraft in which aviator Wiley Post would break all records circling the globe some years later. The Atom was photographed in front of the *Winnie Mae's* prop, then was fixed to the ropes which had been fastened to the undercarriage. He tested the combs and locks and, backing up, towed the aircraft a short distance to judge its rolling weight.

As the crowd watched in silence, the *Winnie Mae* was fired to life. The Atom set himself and yelled "GO!" over a sound like constant gunfire. The pilot gunned the engine, and in the blast of dust and wind and a blaze of searing white pain, the strongman knew he had made a mistake. He had underestimated the power of the big C–6 engine, its water-cooled in-line six cylinders now winding to ever higher revolutions. Lack of adhesion at the wheels was irrelevant, this was "traction" of a more dynamic kind; the Curtiss-Reed propeller which screwed through the air in a blurr of pitched aluminum had the thrust of 150 horsepower behind it. The plane yawed and strained; yet, the Atom held.

At full power, rocking in place, her engine screaming at 1,700 rpm, the *Winnie Mae* relentlessly charged forward. Inside his head he could hear an incessant sound . . . pit . . . pit . . . pit . . . the sound of the scalp being torn from his own head. And still he held. The crowd of startled onlookers shielded their eyes from the tornado of dust.

And then the pilot mercifully throttled down, the engine fell silent. Joe had restrained the airplane for a full minute. Enthusiastic bystanders and photographers converged upon the strongman, as Nussbaum pushed his way through the crowd to help the Atom to one of the airfield offices. He had been hurt, the skin of his face and head nearly disfigured by stress. After massage, the pain died slowly. He vowed never to try anything as foolhardy again.

The next day, in the Ab-Gold office, Joe was depressed because Leah had found out about the airplane stunt and hadn't spoken a word to him since. His management was more enthusiastic.

"We'll get a tour lined up," Goldberg suggested. "Full press coverage . . . newsreels . . . radio. Keep doin' that plane trick, Kid, and you'll be bigger than Houdini ever was. . . ."

"It's not a trick," Joe said, "and I won't do it again."

Goldberg turned to Nussbaum. "What's he doin', Louie, pullin' my chain?"

Nussbaum threw up his hands in an attempt to calm them.

"It almost killed me, Louie," Joe said, as if Nussbaum's eyewitness presence had not been enough.

"What's for free, Kid?" Goldberg insisted. "We're talkin' big bucks here."

"So do it yourself," Joe retorted, ending the argument.

Press coverage of the airplane event had succeeded in moving the Mighty Atom's vaude career into higher gear. Joe progressed from the Brooklyn Frisch circuit to the Mutual, Pantages, Levy, Loew's. R.K.O., and the B. F. Keith's circuit. His take increased accordingly; depending upon the circuit or theater, he grossed up to a never imagined $2,500 a week.

The strongman business was big business, but this elite competition was thinned out by mortality. Strongmen were not long-lived; a strength specialist would eventually burst his heart or blood vessels from the unremitting strain. Even Sandow, the greatest of his time, had died at fifty-eight of a brain hemorrhage.

"You can never know when something inside will tear loose," Joe would say of his chosen profession. "And once the red tide goes out, it doesn't come back for the second show. A strongman is a walking time bomb."

The danger to the Mighty Atom was compounded by the length of his act. The key to survival was, paradoxically, that the most arduous feats had to be accomplished with as little strain as possible. The man who ignored these invisible hazards and turned purple three shows a day to perform his feats would pay dearly for his folly.

"It is the art of holding back even while giving it all you've got," Greenstein said. "But try to explain that to people and they think you're joking."

The Mighty Atom was considered something of an oddity even among the vaudevillians, that crazy quilt of individuals ranging from serious tenors to goldfish swallowers: an exotic zoo of bird and animal acts, comics of every flavor, song-and-dance men, jugglers, musicians who could play anything and everything from a Steinway to a bow saw. Joe sometimes reflected that he had not intended to become a showman, but merely hoped to demonstrate the life-force, and if possible to earn a living. Despite himself, he had become enamored of theatrical life and of the other performers, though often misunderstood by them. Between shows, the Mighty Atom waited in the Green Room, the performers' lounge, where the others played cards or kibbitzed to fritter away their time. Though he enjoyed their company, Joe was usually off in a corner studying one of his ever-present books on anatomy or naturopathy, or polishing the poetry that he had taken to writing. The other players affectionately referred to him as "the Professor."

The waiting between shows was unpleasant for him. Before his first show the makeup session took an hour, the application covering his entire body. If he did less, he'd look like a ghost out in the hot lights. Too troublesome a routine to endure before each performance, whether it was two or five shows a day; once the makeup was on it remained until the last curtain. Greasepaint closed the pores, hampering the skin from breathing, and it was his makeup even more than his act that exhausted him. A still further strain was the traditional bearing required of him. Although a feat of strength had to be obvious in its difficulty, it had to be accomplished with a modicum of grunting, groaning, or visible strain. To keep audience respect the Atom had to maintain stoic, even smiling composure and dignity no matter what the agony.

Joe became convinced that within the Bible, the Talmud, the Kabbalah, there existed the means to a higher plane of life and energy. His formative time with Volanko, his experience with anti-Semitism in the Old and New Worlds, his probings into the self, had crystallized within him a conviction that everything he was and sought to be was as a Jew. If his lot was to appear on the vaudeville stage, then he would present himself as what he was above all, a representative, a son of Israel.

In his strongman's career he had gone through several visual changes, from leopard skin to simple black leotard. Now he collected Biblical illustrations. From these, he had a leather tunic made; the single strap across the chest bore an embossed gold Star of David. He had a headband designed to keep his skull cap in place even while performing feats of strength. Lastly, he re-created Biblical footwear: sandals, with leggings laced to the knee. (The attire was not inappropriate. The philistines in the audience were more than a few.)

He soon learned that out of the impressive sum he was earning there were still daunting bills to pay. He used six onstage assistants; they each earned seventy-five dollars a week. The advance man got $175. The Atom even had to hire his own master of ceremonies at $150 per. Naively, he had hoped that in vaudeville he could communicate his message of doing away with self-imposed limits to create a higher man, but Nussbaum was right; the vaude stage was not the place to do it. Like the rest of show business, vaudeville meant dollars and cents, billing, press agents, and playing to the gang in the balcony. Though he longed to do so, Joe could not share his convictions while doing a silent act after a tap dancer and a dialect comic. He did what was called a "dumb act." One couldn't announce and accomplish feats of strength at the same time, and he dare not speak while actually performing, lest the audience know he was out of breath and the illusion of ease vanish. There was so much he longed to communicate, but he could say nothing.

There were the other "incidentals." Ironically, under union rules the strongman was forbidden to move his own eight pieces of baggage. Whether it was a handbag or a piano, whether one pound or a thousand, every piece of the act's baggage in and out of the theater cost five dollars. Then, there were the obligatory tips to curtain men, spotlight men, stagehands, and others. Such tips had to be paid, otherwise something might "go wrong" during one's act. It was little more than blackmail money. After that came the hotel bills and transportation for the entire act. While the Mighty Atom was earning a small fortune in grosses every week, his net often came out to a loss.

The split weeks, playing two theaters in the same week, were the hardest. A solid week in one location might not insure a packed house every show. Therefore, an agent might play him three days in one theater and three to four days in another, attempting to book seven days a week. Joe could have a day off when he was out of work, though he tried not to violate the Sabbath. Frequently, he drove late into the night to make the next theater, or quietly froze in a dark railroad station waiting for his connection, his sanctuaries a succession of drafty dressing rooms. But there was always the childlike thrill of making people happy, of having them forget their troubles for the price of a ticket. He enjoyed the camaraderie with the other performers, a tightly knit group more like a family than a business. And of course, there were the cheers, the ovations, the glory of being "The Strongest Man in the World. . . ."

"LADEEZ AND GENTLEMEN, I AM ABOUT TO PRESENT TO YOU A MAN OUT OF ANOTHER WORLD. A MAN WHOSE POWER STAGGERS THE IMAGINATION. A COMPARISON WITH THE RECORDS OF THE WORLD'S STRONGEST MEN REVEALS THAT HE HAS NEVER HAD A RIVAL; THAT HIS COLOSSAL STRENGTH BELONGS TO AN ERA BEFORE HISTORY, TO A TIME WHEN OUR FOREFATHERS WRESTLED WITH GIGANTIC DINOSAURS FOR A FOOTHOLD UPON THE SWAMPY EARTH.

"LADEEZ AND GENTLEMEN, IT IS WITH PRIDE AND PLEASURE THAT I PRESENT TO YOU . . . THE ONE . . . THE ONLY . . . *THE MIGHTY ATOM. . . .*"

The applause was euphoria, and he lived for it. To him there could be no thrill quite like it. Not that he wasn't offered others. Louis Nussbaum and the Mighty Atom came to a parting of the ways when, after a few disagreements, the promoter went to California to pursue a business relationship with Mae West. Louie said he was well-acquainted with her, and was titillated by the idea of matching her up with the strongman. Louie's suggestion was less than discreet. "She'll *looove* you, Joe. With that body of hers, and that body of yours . . . *it'll be a clash of titans!*"

The Atom declined. Nussbaum departed for Hollywood, the sunshine sewer, bewildered at such a philosophy. Several years later he surfaced, appearing on the screen in a bit part with "Diamond Lil" herself, in the Mae West film, *She Done Him Wrong.*

Late in 1928, Greenstein's new agent/manager Al Pitroff received a hurried call from a man named Roy E. Van. Van wanted the Atom to come to Buffalo for a week. The bill he had playing in his theater was stinkola and his show business was no business. He needed the best act Pitroff had. Pitroff declared that the Mighty Atom would jam his place, and he was quickly booked as feature attraction at

Buffalo's Gayety Theatre. When he arrived in September of 1928, whatever hasty advance work had been done was to no avail. Business was indeed terrible, as indicated by the standard backstage vaudeville sign: "Don't Send Out Your Laundry Until After The First Show."

In a fit of self-confidence, the Mighty Atom had sent out his laundry *before* the first show. He could not believe the placard could be meant for him. It was the same story as Long Island. Previous disappointing acts had kept the crowds away.

After the first show, Roy E. Van met him offstage. "I told Pitroff, but he wouldn't believe me. It's not your fault, Atom." They went into Van's office.

The Mighty Atom was not about to die in Buffalo. He wracked his brain for some enterprising piece of ballyhoo which could attract sufficient public attention to save the situation.

Van brightened. "Atom, why don't you do that airplane stunt that you did on Long Island. We've got an aerodrome here, and I can get you a plane. . . ."

Greenstein was adamant. "I wouldn't do that again for a million dollars."

"Listen, if you can think of something better than the airplane, I'd like to hear it." There was nothing else. "I don't think it's fair," Van protested. "You did it for those guys on Long Island, and I think I'm paying more than they did. Your booker Pitroff gave me his word . . . *and he gave me your word, too*."

Van was right. He had a right to expect nothing less than what he had been told he would get, a full house every show. He had gotten assurances and paid his money in good faith. The obligation was now all one way. For a moment, the strongman put aside the pain and terror of Curtiss Field. "All right, Roy . . . I guess I can do it again." The words were hardly out of his mouth when he wished he could recall them, but Roy Van was already out the door, setting events into motion.

After his usual evening call, a few minutes of small talk with Leah and the kids, he felt less alone.

"Hello, Pop? When are you coming home?" a background chorus of kids shouted into the phone.

"Is everything all right, Joe?" Leah sensed something wrong from the tone of his voice.

"Business as usual, Mom," he assured her, saying nothing of what the next day would bring. The event would be covered by local publicity; the family would hear nothing of it. He had not lied. It was business as usual. Risking one's life had become part of the strongman game, and if he didn't like the competition, he could always go back to feeding ten kids by scrounging for nickels and dimes.

On the gray, chilly Friday morning of September 28, 1928, Roy Van drove him out to Buffalo airport. The place was swarming with newsmen, photographers, and the usual collection of thrill seekers. His senses had not yet dulled; he had not yet forgotten the pain of that first attempt, or let fade the sound of his scalp being ripped from his own head. He had healed somehow. And now, here he was again. More than likely he would be sent home in two shopping bags.

The car rolled up to the administration building. He could see the plane. His heart sank. It was bigger than the *Winnie Mae*. As the crowd looked on in silence, some stranger made a suitable announcement. "Ladeez and Gemmen, the Mighty Atom . . ." he offered a grandiose introduction. Joe wondered if he would tell them the truth—"Folks, this madman is about to let an airplane tear his head off. . . ."

The Atom met the pilot, a pleasant enough fellow named Frank Little, rather innocuous looking for a probable executioner.

It was a big plane, a new Fairchild FC–2 powered by a nine-cylinder Wright Whirlwind J–5, an engine identical to that with which Lindbergh had flown the Atlantic sixteen months before. With its 9-foot steel propeller, the aircraft could fly at 119 miles per hour. It sat on the runway like a dark winged angel. Two field mechanics unwound the cable attached to the monoplane's undercarriage, to which Greenstein shackled his comb and locks. Solemnly, he waved his "ready" to the pilot, who returned the gesture and climbed into the cockpit. "Contact." A mechanic spun the propeller once, twice, the motor fired, the chocks were pulled from the wheels, and the crowd hung onto their hats from the blast of propwash

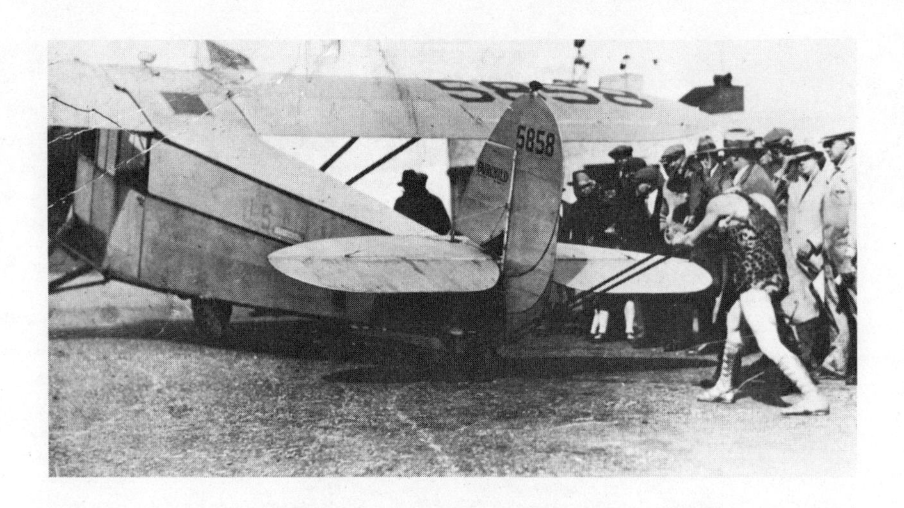

before beating a retreat. Little opened the throttle, the aircraft lurched forward, the cable snapped taut, the strongman was enveloped in the whirlwind.

And then the pain began. That first moment was unadulterated terror. At 2,000 RPM the J–5 developed 220 horsepower. One split second of psychological or physical weakness, and the craft would pull his head off and drag it down the runway.

Little gunned the throttle open to 1,600 RPM, where air speed would have moved past sixty miles per hour. The aircraft was still stationary. And now Greenstein could hear that terrible sound again, pit–pit–pit–the scalp tearing loose. Water poured from his eyes.

In the vortex of the cyclone, every muscle strained to bursting, he engaged in the ultimate battle, that of man against machine: his flesh and bone against the mindless roaring power, he stood his ground. It was the undertaking of a madman. He had survived the *Winnie Mae,* but this winged monster would surely decapitate him. Ten seconds . . . twenty . . . thirty . . . each second swept him further into the riptide of pain.

Like a man on a rocket sled, his face contorted by stress, he fought to hold. Almost like death itself, he was caught in the maelstrom of wind and pain and whirring pitched blade. And then the engine was quiet. It had been the most harrowing experience of his life.

The crowd roared its approval and surrounded him. He had held the airplane back for a full minute. A cloth was hurriedly thrown

BUFFALO EVENING TIMES

SATURDAY, SEPTEMBER 29, 1928

'Mighty Atom,' Super-Strong Man, Pits Brawn Against Plane, Wins

One of the most astounding feats of strength ever seen here was staged at the Buffalo Airport yesterday by the Mighty Atom, who lays claim to being the strongest man in the world, when he held back with his hair, an airplane which had been given "the gun."

The almost unbelievable feat was staged on the runway in front of the airport administration building before a crowd of excited stunt fans that crammed every nook and cranny. So spectacular was the feat that the huge throng of onlookers did not even dare to move while the human dynamo tested his strength against the most powerful vehicle of the age.

Pilot Frank Little was in the cockpit of the plane as the Mighty Atom, adjusting an odd-looking comb to his hair, stepped out of the administration building. To the comb was attached a large cable, the other end of which was fastened to the plane.

The pilot gave the plane speed, turning the propellor for about 800 revolutions a minute for an approximate speed of 30 miles an hour. The Mighty Atom held his ground, seemingly with ease, but when the pilot increased the speed to 1,600 revolutions a minute, for a speed of 60 miles an hour, it seemed the strong man must give away. He held his ground admirably, however, and successfully turned the trick.

The strong man was cheered to the echo when the airplane pilot gave up, but he retired immediately to a private room of the administration building. He was none the worse for his experience. It was announced later that the Mighty Atom would further demonstrate his power between 12 and 2 today by pulling three automobiles, hooked together, around the downtown streets.

Roy E. Van, general manager of the Gayety Theater, one of the interested spectators at the exhibition, sought out the Mighty Atom after the performance, and after congratulating him on his remarkable feat, tendered him a contract. Following the conference, Mr. Van announced that the Mighty Atom will appear at the Gayety Theater all next week, beginning with the Sunday night performance.

over his face to prevent anyone from seeing him and he was ushered into the administration building.

He had been disfigured. There could be little telling if the face had eyes or not. Skin that should have been on the cheeks was lifted almost to the forehead. It didn't even look like a face, the skin had been so contorted by stress. Whether it would be permanent, he didn't know—the greatest terror of all.

A hurried call for ice, packs of it applied to his face and head. "The last time. Never again, a damn fool stunt like that."

Once again, his remarkable powers of recuperation came to the fore. After some hours rest in his hotel room with head and face packed in ice, he was feeling almost human again. Then with proper massage to the wounded areas, his face assumed its normal aspect.

For all the pain, that feat in the Buffalo airport had done the trick, and Roy Van was ecstatic with the response. When the story broke in the newspapers, the Gayety was packed, standing room only.

The Buffalo *Evening Times* had carried the story:

"MIGHTY ATOM, SUPER-STRONG MAN, PITS BRAWN AGAINST PLANE, WINS."

Van actually had to throw open the doors to the theater and send for chairs, so the people outside could stand on them and catch their glimpse of the modern Samson. For his conquest of airplane and sagging box-office, the Mighty Atom was promptly awarded the keys to the city of Buffalo by Mayor Schwab.

The next day, the newly crowned King of Buffalo had just entered his dressing room after the second show. Dripping with sweat and with the roar of the audience still in his ears, he wanted only a good dinner and an hour with his books for a course that he was taking at the American School of Naturopathy. But instead he found a showgirl clad in nothing more than bra and panties lounging on his chaise.

"Hello, Muscles," she cooed at him.

"What do you want?" he asked with startling naiveté. She smiled provocatively. Gently, he lifted her in his arms and deposited her in the hallway. To veterans in the business, it seemed hardly believable that a heterosexual man of his physical attributes should turn down such sexual invitations while on the road. After the incident, and several others, he took to locking his dressing room while he was away. For this, he earned yet another nickname, Padlock Joe. In the business, he achieved equal notoriety for these backstage "feats of strength."

At the conclusion of the engagement at the Gayety Roy Van escorted him to his taxi. "Atom," he said, "I can't figure you out. You run away from women; you make money and don't count it; you headline the bill and you own one suit. For chrissake, Atom, buy yourself another suit!" He pressed an envelope into the little strongman's hand; inside was a bonus of five hundred dollars.

A short time later, the word came down to the New York airports that the airplane stunt the Mighty Atom had performed had been banned by the authorities. Greenstein was relieved. He had recommended in interviews that no one ever try it. He stated publicly that he would never attempt it again.

Sometime before, while he was appearing in New Jersey, a boy named Louis had come from Paterson to see him. He came backstage, a powerful young fellow with a shock of red hair and lofty ambitions; he too wanted to be a strongman. Just as Greenstein had duplicated the act of the Great Kronas years before, Louis had his mind set to duplicate the Mighty Atom, and eventually to surpass him. Greenstein tried to discourage the boy from attempting a professional career. The strongman business wasn't all it appeared to be; the glamour was veneer, the risks were many, the work was a grind. But the young man was not to be dissuaded.

Louis came to see him on the vaudeville trail. He was all fired up, eager to duplicate the Atom's airplane feat. Joe begged him not to try it.

"You're playing with dynamite, kid. It almost killed me."

"Yeah, sure . . ."

"It's illegal in New York."

"I'm going to do it in California. It's not illegal there yet."

He was a powerful kid, about twenty-three, built like Jack Dempsey. The Atom told him in detail how dangerous the airplane stunt was, but he insisted. When all else failed, Joe wished him well.

A month later, word came back to the Mighty Atom. The kid had indeed tried the airplane number. But instead of just attempting to hold his ground, he had rigged up a post on the runway and tied himself to it. When the aircraft revved and charged forward, it tore his scalp out. He bled to death right there on the runway.

VAUDE NIX, PIX CLIX

From Buffalo, the Mighty Atom went on to play Niagara Falls, Syracuse, and Rochester. More sellout business, and the keys to those cities from Mayors Loughlin, Hanna, and Wilson. In Syracuse, he had played the Quirk Theatre, where in addition to the bill of vaudeville was an extra added attraction: the film *Riders of the Dark,* starring Tim McCoy. In Rochester, before leaving town, the Atom towed a fully equipped and manned hook-and-ladder fire truck up an incline.

The day after Election Day, 1928, found him appearing at the Rochester Family Theatre in Rochester, New York. Business and reviews had been excellent. Having already done several shows, he stepped into the theater alley for a breath of air before the last performance. Someone slapped a newspaper into his hand; the headline glared at him: "HOOVER ELECTED PRESIDENT." Joe Greenstein disliked Herbert Hoover. Angry at the election results, he barely got back into the theater for the third bell, the backstage signal for performers to be onstage. He rushed in, took off his cape and went to work. For one of his first feats of strength, he would with a bare hand, cushioned only by a thin handkerchief, slam a twenty-penny spike through thirty plies of galvanized sheet tin atop a solid 2-inch pine board.

"I didn't have my mind on the act. I was thinking about that bum,

Hoover. . . ." He wound up and slammed the nail right through his hand. He had gripped it by the wrong end with head down. The spike went clean through. Oddly, there was little flow of blood. The act was halted as his two male assistants pulled it out. He treated and sterilized the wound himself, bandaged it, and went on with the act. After that occurrence, he always made a practice of examining the nail three times before trying the stunt.

Returning home, the Atom was invited to perform at Bryant Hall, Forty-first Street and Sixth Avenue, at a show given by Siegmund Klein, physical culturist, trainer of champions, owner of Klein's Gym in New York City, for an audience composed almost entirely of weight lifters, strongmen, and sundry giants of assorted sizes, shapes, and pursuits. Most of them outweighed Greenstein by one hundred pounds or more.

For his last number he would while standing erect and looking up at the ceiling have four men, two on either side of him, bend a heavy bar of cold-rolled steel down over the bridge of his nose. The difficulty of this feat can be demonstrated by a party trick, whereby a grown man can be stopped from moving forward by the back of a single finger pressed against that spot.

To hold the bar of steel in place, Greenstein had devised a small formfit aluminum nosepiece on which the steel would rest. Now, before the audience of his peers, when the moment came to bend the bar he gave the signal, and one of the four men pulled prematurely, yanking the bar down; the nosepiece slipped and the bar crushed it into his cheekbone. He began to hemorrhage, as he fell back, caught by the next act waiting in the wings. The curtain was rung down.

"Get that curtain up," he ordered, then went out and placed the bar across his nose without benefit of the aluminum device. He gave the signal again, all the while hemorrhaging. Several of the muscle men nearly passed out, but the Atom ignored his condition. When the bar lay crushed on the stage, he strode off.

About a month or so later, the Atom was invited as a guest to New York's Pioneer Gym, 338 West Forty-fourth Street, a popular training ground for prize-ring champions as well as hopeful contenders and amateurs. Greenstein took along his friend Benny Platt, a wry, cherubic little man.

The Atom was introduced to two heavyweight professionals, Jack Renault and Arthur DeKuh. Renault was the Heavyweight Champion of Canada, and in seventy-five professional bouts had fought everyone from Harry Greb to Jack Dempsey. DeKuh was a boxer with a similarly impressive record. For a publicity picture, both fighters were invited to try to pull the Mighty Atom's hair out. All present treated it as a gag, but the attempt was going to be made in earnest. As a photographer readied his camera, the little Samson parted his shoulder-length locks down the middle; Renault on the left side, DeKuh on his right, each man seizing handfuls and readying his massive arms and shoulders. The Atom told them to proceed, and both men strained and groaned for real; but to no avail. The Atom's hair could not be pulled out. He was indeed, as Ripley had dubbed him, "The Man with the Iron Hair."

It was a bizarre demonstration and cause for much good-natured

clowning. One of the fighters in the gym, a known veteran contender of the heavyweight class, scoffed at the demonstration as mere "showbiz." That it obviously was, but the boxer was rude and abrasive in his comments.

"That little guy's nothing special," he said to Benny Platt.

"That guy's dynamite!" Platt defended his friend.

"One shot and I'd knock him down like a fly," the fighter replied.

"You couldn't knock him down at all! He could put you flat on your can, *and I've got fifty bucks that says he can!*"

The Atom looked on with faint amusement.

The Heavyweight said he would accept any contest that Platt cared to propose. Platt suggested that the prizefighter take a free shot at the Atom's unguarded midsection. If the strongman held his ground, he would win the fifty dollars.

Greenstein stepped up and took his stance, the same one he used while holding back airplanes. The boxer wound up and let fly, his gloved hand sailing into the smaller man's stomach. The smiling Atom hadn't budged.

Platt laughed in the pugilist's face. "Care to try again for a hundred? Same thing." Accepted. The fighter's fist crashed again, this punch stronger. But still, the little man could not be moved.

Platt goaded the boxer on. "Take one last shot at him for another hundred bucks. If you can move him, we'll give you all the money back." Enraged by now, the fighter set himself and delivered a vicious hook to the Atom's midsection. The little man wavered for a moment, but kept his stance.

"Okay, I'll give you a chance to keep your money," said Platt. "Give us a turn. Let the Atom hit you one on the jaw. If you can keep your feet, you won't owe us a thing." Confident, the pugilist stuck his chiseled jaw forward. Thwock! The Atom hit him a clean shot, his knees buckled, and he collapsed to the floor.

On the way out of the gym, Greenstein had a feeling that it was more than mere chance. His friend, now counting his money, seemed to know what he was doing. Platt, a prizefight aficionado, knew the boxer's record. "I knew he had a glass jaw," Platt smiled. "And a glass jaw is never fit company for a big mouth."

- - -

If the pugilist had a crystal jaw, the Mighty Atom certainly did not, performing his biting numbers at almost every show. He made a habit of giving out souvenirs, and in restaurants would crush a half dollar in his jaw and leave it as an extra tip. The coins that he destroyed in the act were sometimes his own, and sometimes those offered to him from the audience.

While working the Rochester Family Theatre, he had just deformed a half dollar in the usual unusual manner and tossed it to a child in the front row, when a man rose from the audience in the midst of his performance and flashed a badge.

"Mr. Atom, you've got a wonderful act, but I can take you off the stage right now, and get you as much as five years in prison."

He flashed the badge again for emphasis. "Treasury Department. You're destroying U.S. Government currency." The Atom blanched. "But . . ." The Treasury man added, "If you promise now in front of all these people to mend your ways, I won't trouble you further."

The strongman quickly raised his right hand. "I give you my word, sir. I'll never do it again." And he didn't. The jaws of steel would stick to other paraphernalia; the Treasury Department didn't concern itself with horseshoes or tire chains.

Of course there were some people who apparently could never be convinced. While working vaudeville near Paterson, New Jersey, he was in the midst of the day's second show, standing in the spotlight as the music and drum roll swelled. He held up a heavy length of chain, placed a link between his back teeth, and bit down. Pop! He held up the sheared ends, the chain in two. As the applause subsided, a voice angrily called out from the audience, "Fake! Fake!" The house fell silent, awaiting the strongman's response. Joe stopped his act and, shielding his eyes, peered out over the footlights. "Who said that?"

"I did." A man stood up. "My name is Doctor G———. I'm a dentist and I say you are a charlatan. What you claim is impossible to do. Human teeth cannot bite through iron."

"Mine can, and do," Joe replied to the disgruntled dentist.

"Nonsense. A cheap theatrical trick. I challenge you to come to my office for an examination. Then I dare you to do it again in my presence."

"Sir, your cynicism is understandable, but I assure you that these feats are legitimate. What I do is strictly on the level."

"Nevertheless," the dentist replied, "an examination in my office will prove who is right."

The next morning, bright and early, the Mighty Atom strolled into the dentist's office. Dr. G——— was all business and motioned him to the examination chair.

"Open wide, please. . . ." The adjustable light shone in Joe's eyes; the dentist picked up the tiny mirror affixed to its surgical steel handle and began to probe.

"Hmmm."

"Whaddya thee, Doc?"

"Well, I don't see steel caps or appliances of any kind," he said with disappointment, and still suspicious, continued poking around. "These teeth are close to the gum line, but aside from a few chips here and there, they appear perfectly normal." He continued his search, all the while alluding to the Mighty Atom's lack of veracity. After several frustrating minutes, the dentist was somewhat disconcerted.

"Finithed, Doc?" the Atom inquired with the steel probe still in his mouth.

"Just about," the crestfallen dentist replied.

Abruptly, the strongman closed his jaws. Crack! The surgical steel arm of the mirror was cleaved in two. The Atom casually rose from the chair. . . . "Ptooey!" He spat the severed end of the steel mirror onto the floor, winked at the startled dentist, and closed the door behind him.

A short time later, Dr. G———'s signature appeared on a document along with a dozen other men of his profession attesting to the Mighty Atom's supernormal ability to bite through iron.

The Mighty Atom was largely unaffected by his surroundings, and it made little difference to him where he worked. He concerned himself with what he did, not where he did it. The Hippodrome on Sixth Avenue between Forty-third and Forty-fourth streets was the

colossus of vaude—a mammoth place with a seating capacity of nearly five thousand, a stage like a football field, and dressing rooms named after the states of the Union. The sound of applause at the Hippodrome was like an ocean's pounding in, wave after wave. He worked the Palace, the temple of vaudeville and the dream of every performer.

If he was happy with his performance, he would just as soon do it at the New Delancy, on New York's Lower East Side. It was much like other small-time houses that catered to immigrants; programs were printed with English on one side and Italian or Yiddish on the other. He worked Harlem's Apollo Theatre for an enthusiastic black audience. Wherever he worked, big- or small-time, he did his turn and took what benefits were to be had.

Managers, agents, and anyone who was involved at the moment were allowed to care for the "details" and finances. Others collected his money, paid the bills, and gave him what was left. In the transition, untold thousands of dollars evaporated. So much for the cliché of the business-minded Jew; Greenstein was the worst businessman in the world. He had, almost unconsciously, adopted the study habits and life-style of his rabbinical forebears, and he disregarded material matters. The Mighty Atom had not a scintilla of accountant in him; he suffered the consequences.

During those Depression years as before, it was Leah who created a buffer between the empty vanities of showbiz and the family. It was she who made the Greenstein home a shelter where all were welcome day or night. The clan had grown to ten with the birth of their last child, Jerry, born one month before the crash of Black

Courtesy *Philadelphia Inquirer*

Andre LaTerza

Lying on a board studded with sharp-pointed nails, Greenstein supports 17 members of a band and their instruments to entertain crowds at New York's Coney Island. The total weight of the musicians and platform is well over a ton. The "gigantic dwarf" has to remain in this position 15 minutes while the people are mounting and leaving the platform.

THE PHILADELPHIA INQUIRER, MAY 21, 1939

Tuesday in October of 'twenty-nine. It seemed that Leah was always surrounded by children, and often as not they were someone else's. She was the proprietor of a nonprofit playhouse and candy store where it was understood "We never close." The kitchen was the nucleus of the house. Joe wondered why food didn't spoil, because the ice-box door was always open with a kid standing in front of it. Leah kept a cauldron of soup simmering on the stove at all times for friends, relatives, down-at-the-heels vaudevillians or any hard-luck case who happened by. Sometimes the Greenstein home would cheerfully accommodate another whole family who had fallen on hard times.

Joe had no favorites among his children, but the one who most resembled him was little Mendel, his only blond child, who would sit on his lap for hours to be read stories or just sit quietly.

Despite her own burdens, Leah went out of her way to make him feel a success in the eyes of his children. When she baked for the family, he would put on an apron and help in the kitchen. And though she had done all the work, if he so much as stirred something in a bowl, she would serve it to the family and exclaim, "Look what Pop made!"

Though certain of his unique abilities, Leah recognized that his risks on the bed of nails were increasing. As the Atom continued escalating his feats atop the spiked board, serious injury became inevitable. Leah understood that he could never limit himself. Quietly, after one of his more harrowing local performances, she took an axe and chopped up his bed of nails.

He made another one.

As the 1920s drew to a close, Louis Loomus and Bert Jonas, who had booked him for the United and Pantages circuits, scheduled the Mighty Atom on a tour of the Continent. In addition to his assistants for the act, the Atom was accompanied by his manager and friend Al Pitroff. A kindly, sad-faced man with an air of dignity, Pitroff bore the lofty title, "Mystery Master and Handcuff King." He served as master of ceremonies during the act, and it was his responsibility to keep things running smoothly as road manager. As usual, the Atom was single-minded: he didn't want to have anything to do with details. He would simply come onstage, do his turn, and retire to his hotel room for a few hours' rest. As they made their way through England, France, Belgium, the strongman frequently didn't know what country he was in, let alone which theater he was playing.

He had agreed to go on the tour because vaude was getting tougher by the day. American variety entertainment was changing; instead of the usual two a day, he was working four shows a day in the lesser theaters, and even these dates were getting scarce. In the cities of America, bread lines were longer than ticket lines. He hoped that the change of a European tour might present new prospects. No such luck. The money situation did not improve, and it was the same old story: by the time he paid everyone, he could not pay himself.

The Golden Age of Strongmen had long since passed. The immortal men of iron like Sandow, Saxon, Cyr, "Apollon," and Rolandow all had reached fame in the 1890s when strength acts were at their zenith. The Continent had seen the best of them, and a roll call of the great strongmen sadly showed that this little American, while impressive, had been born too late. The last popular performer had been Siegmund Breitbart, and after his death, the strongman business was somewhat passé in Europe. Lukewarm audiences combined with his homesickness and money problems. He was glad when he finished the European tour. To him, Europe was tired and beaten.

When he returned to America, he could see that live vaudeville was doomed by mechanized entertainment. Vaudeville was a business of "specialists" yet each act, with a style and offering of its own, was integrated into a totality of entertainment; it was the medium that had become antique. The Mighty Atom had arrived during its final vogue and the stay had been all too brief.

The Depression, the electronic wonder radio, and talking pictures had taken their toll of live American variety theater, but talking pictures landed the fatal blow. The short-reel silents had begun as "chasers," an extra added attraction; the dessert had stayed to consume the dinner. The end would come in 1932 when the New York Palace, the shrine of vaude, would cease its big-time "two-a-day" policy and become a "five-a-day" presentation house.

With vaude houses closing daily, the Mighty Atom filtered back down to the small-time, and then to burlesque. Vaudeville had been a family show, suitable for women and children. Burlesque was a show of orifice humor. He hated it. The women were mostly "bums," he declared. He worked the burlesque circuit for a short time, out of necessity, before calling it quits.

Years before, Ringling himself had offered him a job with the Ringling Brothers, Barnum and Bailey Circus, but this, too, he had declined. Vaude and vaude alone was in his blood. The vaudevillians he knew and worked with had been warm-hearted and generous to a fault. Now those who had been headliners wandered the stone canyons of Broadway with photos and clippings under their arms, looking for a place and time that no longer existed. Like birds in flight with no place to land, they searched for work; but there was no work, and never would be again.

New York City was the vaude capital of the world, but on every street and booking office it was the same. "Nothing today." The Mighty Atom, who had earned thousands a week at his zenith, now did not have a nickel to get on the subway. He walked from Brooklyn to Manhattan several times a week looking for work. On September 18, 1931, after a day of pounding the pavements, he returned home. Before he walked through the door he could hear the volume of sounds inside; the place was filled with neighbors, the kids were crying or sitting stunned in corners of the room.

"Pop . . . Pop," Leah whimpered at the sight of him.

"Leah, are the kids all right?" His eyes swept the room. It was the first thing that entered his mind. All the children were home, except Mendel.

"Mom, where's Mendel?"

"He's dead . . . ," she shrieked.

The boy had been struck and killed by a truck while playing in the street. The family plunged into chaos. After the traditional period of mourning, Joe's despair remained. The love of his other children could never assuage the loss of his Mendel, his little blond pal, upon whom he lavished so much affection. Never before had Joe Greenstein been made to bow to circumstance; when there was no money, he made what was needed with his hands; when the kids were sick, he healed them. There was nothing that could not be done, no situation from which he and Leah could not recover, no hurt that could not be made well again . . . until this. They had lost a measure of themselves. She who had always managed a smile in the midst of any hardship now abandoned her tradition of singing to the children in the evening. After the death of her eighth child, Leah did not sing again.

THE PITCHMAN

Joe Greenstein scoured the streets of New York looking for work. The managers and agents who had previously negotiated four figures a week had evaporated almost as quickly as they had appeared. It seemed the world no longer had an interest in strongmen. There were a few left, but relegated mostly to dime museums, the dregs of the amusement business.

Finally, he hit bottom, working Hubert's Museum and Flea Circus on Forty-second Street. The Mighty Atom was in good company. When Jack Johnson hit the skids, he too found himself at Hubert's, appearing for a five-cent admission. The Mighty Atom performed eleven times a day. When the constant stress without rest nearly killed him, he called it quits.

Aimless, he wandered up and down Broadway. He cursed himself for his character flaws, for his lack of business acumen. He stood with hundreds of other Depression unemployed in front of the stores operated by pitchmen. These were a flying circus of disbarred lawyers, mad doctors, and assorted desperate characters; the gypsies of Broadway, they rented by the week and hustled elixirs and magic powders guaranteed to restore one's hair, or relieve any condition from warts to terminal pneumonia. An "Indian" with a classically Italian face and a Brooklyn accent stood replete with feathered headdress and harangued the crowd. He did his Brooklyn war dance as he peddled snake oil. The Atom surmised that he had never been west of Hoboken.

The next man up was a "professor from Heidelberg." A pitchman would do his fifteen-minute spiel, then hustle the timmies and get off, allowing the next independent contractor to mount the platform and do his bit. It amused Greenstein. How could people believe such crap? These "authorities" on health, nutrition, and healing were obvious quacks. Yet people believed and bought.

Over the years, the Mighty Atom had acquired an immense knowledge of physical culture and naturopathy. Shortly after arriving in New York, he had begun studying under Dr. Benedict Lust at the American School of Naturopathy and Chiropractic on Lexington Avenue. Lust was a medical doctor from Germany who had come to America about the turn of the century and had founded this school which devoted itself to healing through natural means. The naturopath believed that sickness and human ailments resulted from an organism's violation of natural laws. Therefore, preventive measures as well as cures could be gained through natural means: rest, diet, proper exercise, sunshine, fresh air, water, herbology, and chiropractic. As the Life President of the American Naturopathic Association, Benedict Lust had substantial credits; and among his accomplishments was the creation of the first real health-food store in America.

For years the Mighty Atom had pursued these studies, and he had a voluminous knowledge of physical culture, chiropractic, food science and dietetics, chemistry, anatomy, and allied subjects, and had received a certificate as a Doctor of Naturopathy. This complemented his lifelong study of the Mosaic Dietary Code, the Kosher laws.

Although sometimes amused by a skillful pitchman's performance, the Atom was angered by their purposeful and dangerous misleading of the public. Show business was finished for him, his awards and accolades useless. This man who thrived on activity, on lofty endeavor, now lay in bed, whipped, emotionally exhausted. Sometimes he wrote poetry as a consolation. His poems, written in Yiddish, had occasionally been published in some of New York's Jewish newspapers; he had never tried to translate them into

English, for he felt they were untranslatable. He kept a thousand unpublished poems in an old leather briefcase. Now even his writing gave him no solace, and he lapsed into self-pity, a trait he abhorred. But the feeling was still there, the "calling" that pervaded the very air around him. "Show me . . . ," he prayed while sitting in his darkened bedroom.

Leah entered the room. "Yosselle"—she took his hand—"look what you look like."

"Leah, I don't know what to do. I can't even earn a living." He had turned the energy inward, on himself. "I feel like I'm drowning."

"Yosselle, remember what you said . . . 'all for a reason,' " she said, refusing to share his depression. "Everything will be all right, you'll see."

He felt her strength and believed what she told him. Only Leah could make him believe. After dinner he put on his clean clothes, prayed, and went out again.

On Manhattan's Sixth Avenue he stopped at a storefront. A man was selling stretch exercisers. The Atom watched with interest as he happened to know that this one was not a fraud like the other pitchmen, but an athlete with the body of an Adonis and a genuine knowledge of health and fitness. The community of professional wrestlers and strongmen was a small one; Abe Bosches and the Mighty Atom had known one another by reputation. Each had been a professional wrestler; they had the same ethnic background and bore a startling resemblance to each other. Bosches, like Joe a pocket Hercules, had fallen from glory and was forced by the Depression to labor four hours a night as a masseur at "the shvitz," the Luxor Baths, and the rest of the time to operate as a pitchman.

During his speech, he introduced the Atom to the crowd with kind remarks and obvious respect, and after the spiel in which Bosches sold one of his exercisers for a dollar, Joe went over and extended his hand.

"Thanks for the words, friend."

"What are you doing around here, Atom?"

"To be honest with you, I'm looking for work."

Bosches was sympathetic. "You know, I could sure use a hand here."

"Doing what?"

"The average guy will stay in shape using one of these for about five minutes a day." He held up the heavy rubber apparatus that he had been demonstrating. "You know how long I've worked this one so far today? Eight hours. My butt is dragging. You could come up on the platform and help me sell, and I'll split the profits with you."

Bosches was a man who knew talent. Some years before he had hired another young man to demonstrate his exercisers in a Manhattan storefront window. This protégé performed his job very well, and later achieved some fame of his own—as Charles Atlas.

The idea of addressing a crowd filled the strongman with dread. He had performed before thousands as a professional entertainer, but having never had to speak, his shyness had not been a drawback. He had been a health lecturer for many years, but his instruction had always been to individuals. There was no problem in speaking to people one at a time; but speak to a crowd? Sell? It was a humiliation for a man of his previous stature to be relegated to a pitchman's podium.

"I appreciate the offer, Abe, but I wouldn't have the nerve to speak to a crowd. I know what I might want to say, but . . ."

The truth was that vaudeville was gone, he was broke, and the Depression had forced men of higher status to lower ebb. There was no point in deluding himself; he had nothing to lose. He stepped up onto the speaker's platform and began. "Ladies and gentlemen, I . . ." He looked down at his audience composed of two little street urchins who had come in to get warm. "I would like to talk about . . ." The people on the street looked right through him and kept walking. "Ladies and . . ." He quickly stepped down from the platform and hurried out of the place. Bosches thought he had quit from embarrassment, but a minute later the Atom reappeared; in his hand was a fistful of iron spikes that he had bought with his last change at a hardware store around the corner. He began wrapping a spike in a handkerchief. People stopped out of simple curiosity. He began again. "For my lecture I would like to talk about"—he made a quick choice of subjects—"sugar." Bosches looked quizzically at him.

"The average American begins his day with a breakfast of half a grapefruit which is a little too sour for him. So, he puts a teaspoon of granulated sugar on it. He has some cereal whose nutrient and roughage value is approximately that of the box it came in, with more sugar on it. He has a slice or two of white toast, dead starch which becomes sugar in digestion, with jam or jelly loaded with more sugar, a cup of coffee which is caffeine with still more sugar . . . and so on.

"I have only mentioned for example the sugar that he adds with his own hand. This is not to mention the products that he eats which have been manufactured with a deluge that he never sees.

"If the ingredients of this meal vary, the thinking behind it is almost universal. The individual believes he has eaten a healthful, balanced meal, when in fact he has insured that he will be sickly, catch cold, lack energy, be constipated, probably be given to depression, and increase the possibility of suffering or dying from a degenerative disease. Americans are killing themselves at the table, digging their own graves with a knife and fork."

"What's he selling?" one man in the audience inquired of another. "You got me," the other shrugged. After the first half dozen passers-by had stopped to listen, the crowd began to build. To keep their interest, the Atom casually folded the spike in half as if it were tissue paper, and tossed it to one of the children up front. He slowly wrapped another spike as the people watched, now with fascination. "Sugar is not a food. To put it bluntly, it is a death food. It has no vitamins. It will sustain neither life nor well-being. The opposite is true. Sugar has become a drug. It's habit forming.

"Yes, sugar will give a short burst of energy. But what kind? The wrong kind. It will quickly fade. The end of it is a burn-out, a puff of smoke, like gasoline sprayed on a fire. This 'energy' produces no stamina, but a feeling of exhaustion at midday.

"In a given year, most of you listening to me will consume about a hundred pounds of sugar, some individuals ingesting as much as their own body weight. In terms of saturated use, sugar is a relatively modern substance. The human body was neither developed for, nor is ready now to cope with the massive infusion of it.

"Such consumption may directly contribute to the incidence of nervous disorders, hardening of the arteries, coronary thrombosis, diabetes, rotting teeth and gums. Sugar consumption will aggravate the effects of other diseases. It is a degenerative to the human body.

"We are slaves to sugar, and in turn we enslave our children at birth. Examine the label of contents of most prepared commercial baby foods. You will find sugar as one of the ingredients. Walk in any grocery store and look at the children's foods, which have adorable names and are shaped like little animals or other objects of entertainment. There, the amusement ends; these products are heavily laced with the white poison. Once having acquired a taste for them, we consume these anti-health products for life. If only we had stopped to think about the results of our consumption."

A man in the crowd looked at the candy bar he was munching as if for the first time, and with newfound disgust, tossed it into a garbage pail.

"Now why have I spoken to you about this subject? Because frankly it bothers me that you're destroying yourselves with blind habits." He held up the exerciser. "And what does this thing have to do with what I've been speaking about? Everything. Diet and routine exercise are interrelated in terms of general health." He had no qualms about selling the article because he had used a similar one to good effect for years. He spoke another minute on the use and benefits of the exerciser, and after selling three of them, he stepped down. Bosches was nearly dumbstruck.

"You never did this before?"

"Only to individuals . . . Abe, there was so much more I wanted to say."

"Go back up in a few minutes," Bosches suggested.

Joe's blood was up, almost like the intermission between wrestling falls or feats of strength. He knew what he had failed to articulate, but here it could be rectified almost immediately. The Atom retired behind the platform to collect his thoughts. When he returned to speak, his presentation was more coherent, the words even more assured, the lecture growing in authority with every sentence. The crowd now stood ten deep. Bosches watched in amazement: a man who had been afraid to open his mouth only hours before was now speaking in thirty-minute lectures.

Again he returned to the platform with a different subject: fasting.

"Most people have the idea that the more they eat, the stronger they are going to be. That's very wrong. The stomach should hold about a pint to a pint and a half of food. Stuffing oneself to excess causes the stomach to labor under improper conditions. Actually, your life depends more upon what you don't eat than what you do.

"You should only eat when you are hungry. And you are only really hungry when a piece of plain bread tastes to you like a steak. The vast majority of people lead lives that are ruled by the clock. They eat breakfast and before it is properly digested, they are already sitting down to lunch. Before that is eliminated they are having dinner. Before the body has had time to expel that, they are sitting down to breakfast again. Why? They weren't really hungry. They ate because the clock told them to do so, never realizing that their heart and organs were laboring to digest, utilize, and eliminate much of what they didn't need in the first place.

"We never give ourselves a break . . . until we break. Human beings literally eat themselves to death. If only the average man knew that his miraculous body could be made to last twice as long if it were not so senselessly abused.

"Years ago, a man by the name of Volanko taught me that *fasting is the greatest cure on earth, a foundation of internal cleanliness, a preventative against ailments and diseases of all kinds.* I began the practice of fasting in my teens, and I can say with complete assurance that *correct fasting is the key to longevity.*

"This is not a new idea. According to Hebrew tradition, 'On the second and fifth day, thou shalt not eat.' Now, what does this mean? On Monday and Thursday, you fast. This Jewish custom of fasting on those days, when the Bible is read in the synagogue, is said to put one into a clearer, more exalted state of mind. In the Bible, Exodus thirty-four, it is written that Moses fasted for forty days and nights before receiving the Law on Mount Sinai.

"In addition to the traditional Mondays and Thursdays, there are no less than six other fast days in the Hebrew calendar. Religious considerations aside, there is great wisdom to this for its health benefits. The practice of fasting is also employed by Yoga and other Eastern philosophies, as well as by individuals from all walks of life. The beneficial effects of correct fasting are universal.

"Fasting cleanses the human system. When you fast, you allow all of the bodily organs to rest, repair, cleanse, and rejuvenate themselves; what physicians would term 'physiological rest.' Your blood purifies itself; the entire system dispels accumulated toxins and revitalizes. Most importantly, you are giving your heart a chance to rest. When your stomach and heart are in sound shape, the other organs will follow. Fasting will improve or help cure a whole range of ailments from high blood pressure to tuberculosis, and ulcers.

"During this abstention from food, one should continue with moderate activity, work, or school. *It is of paramount importance that during any fast, one should drink sufficient quantities of pure water (spring water, if possible), or fresh fruit juices to prevent dehydration.*

"As a curative, fasting is remarkable; as a preventative of disease, it is indispensable. The human body is miraculous beyond human comprehension for its powers of repair and recuperation. But one must aid the process. Fasting is one of the foremost ways. You are, in effect, allowing your internal organs to cooperate with nature. I have fasted for as long as eighteen days in succession, although I would certainly not recommend this to you. I suggest a simple fast of one day a week; it will increase energy, vitality, mental alertness, and promote a sense of well-being." Back to the exercisers; he sold another ten.

With a slap on the back from Bosches, and a ten-dollar bill in his pocket, he took the subway home. It was as if now, magically, everything he had ever learned or accomplished could be communicated to other human beings. No longer the silent strongman, he could weave his convoluted lessons, discoursing on anatomy, dietetics, kosher laws, weight training, the life-force. He could do these things, and earn an honest living at the same time. Why did the name "pitchman" have to suggest dishonesty and misinformation? He would tell the truth.

He arrived home late that night, kissed Leah awake, and pressed the bill into her hand. "Leah, I know what I'm going to do now. I'm going to be a pitchman." Her expression fell. "No, it's not what you think. I'm going to show people how to improve their minds and bodies. I'm going to give back a little bit of what I've received. I don't need a fancy theater, and no one has to buy a ticket. I can lecture in a storefront, anywhere. I proved that tonight. Mom, my gifts were never meant to be an end, but the means to an end. I know that now."

Though earnings would be less than vaudeville, here he would be making real money. Money in his pocket. Even his inability as a businessman would be overcome. Transactions would be simple: no contracts, no involved arrangements with intermediaries. He would deal directly with people. But what to sell? Abe's exercisers were all right, but not exactly what he had in mind. For years he had considered the idea of having various health products, based on his knowledge of naturopathy created for his own use. Why could he not produce such products in quantity?

He decided to create the first of the Mighty Atom Health Products, a soap. "Why do I sell soap?" he said while crushing horseshoes and dispensing them to a crowd a few weeks later. "Because most of you people don't even know how to wash yourselves correctly. I have created my own soap based upon my abhorrence of the mass-produced stuff." He held up a store-bought bar. "The average article of this ilk is disgusting; a combination of rendered animal fats and lye or other caustic, and perfume to cover everything up. When you rub it into your skin it smells wonderful, and when the perfume wears off, your pores are left clogged with garbage. The messy ring around the bathtub isn't what came off, but the residue of what is going on. When the average person takes a bath, he is not alone. He is taking a few cows, sheep, and who knows what else into the bathtub with him. That ugly smell of human sweat is more the skin's discharging of this animal fat and caustic than the body's natural odor.

"My own formula, which is neither new nor complex, is a natural soap which will not dry out the skin as will those composed of the animal product. This is pure coconut oil and natural lemon."

This was a healthful and economical product. Not only was it long-lasting, but it doubled as an excellent shampoo. The Manhat-

What Sterling Means to Silver
Atom's Supreme Soap means on the Skin

ATOM'S GOLD PIECE

Atom's Supreme Soap
Made From
Coconut Oil
FOR THE BATH
Toilet & Complexion

PRICE 50 Cents

IN QUALITY IT STANDS ALONE
— AND —
Quality Is Our Motto
A Perfect Skin Cleanser
Wonderful for Shampooing. Cleanses the Scalp and removes dandruff. Leaves the hair SOFT, FLUFFY, and LUSTROUS
PRICE 50 Cents
KING OF THEM ALL

International Health Products Co.,
P.O. Box 432, Red Hill, Pa. 18076

SUPERIOR LINIMENT
As an aid in relief of discomforts of rheumatic type pains, muscular aches and pains, and simple neuralgia when due to cold, exposure, or fatigue.

DIRECTIONS
Rub over affected area twice a day.

CAUTION
Do not use on broken skin, near eyes, or mucous membrane.

For External Use Only
2 Fl. Ozs.

Active Ingredients
Crystals of Menthol, Artificial Oil Mustard, Oil Camphor, Eucalyptus, Oil Ceder, Kerosene, Wintergreen, Turpentine Oil, Linseed Oil, Origanum.

Send us two dollar and we will send you a 4 oz. bottle of Atom's Wonderful Superior Liniment.

INTERNATIONAL HEALTH PRODUCTS CO.
121 EAST 96th STREET
Brooklyn 12, N. Y.

A UNIQUE PRODUCT – WHAT STERLING MEANS TO SILVER – ATOM'S SUPERIOR LEMON COCO OIL CASTILE SOAP MEANS TO THE SKIN

ATOM'S GOLD PIECE

Atom's Superior Soap
Made From
Lemon Coconut Oil, Castile
For The Bath
Toilet and Complexion
PRICE 50 CENTS

IN QUALITY IT STANDS ALONE
— AND —
QUALITY IS OUR MOTTO
A PERFECT SKIN CLEANSER
Wonderful for Shampooing. Cleanses the Scalp and removes Dandruff. Leaves the hair SOFT, FLUFFY and LUSTROUS.
PRICE 50 CENTS
QUEEN OF THEM ALL

SPECIAL OFFER
Send us $2.00 and we will send you
8 Cakes of Atom's Superior Castile Soap
25 Cakes for $5.00

CONTENTS

Sassafras Root	Althea Root
Hungarian Chamomile Flower	
Caraway Seed	Alfalfa
T. V. Senna	Berberis Aquit
Chicory Root	Licorice
Peppermint Leaves	Coriander Seed
Golden Seed Root	Spanish Aniseed

BEWARE OF FAULTY ELIMINATION
HELP TO AVOID THE FEELING OF TIREDNESS THAT MAY BE DUE TO IMPROPER ELIMINATION.
LAXATIVE AND CARMINATIVE
Improper Elimination is often the cause of Sickness

KEEP THIS CLEAN

For Temporary Constipation

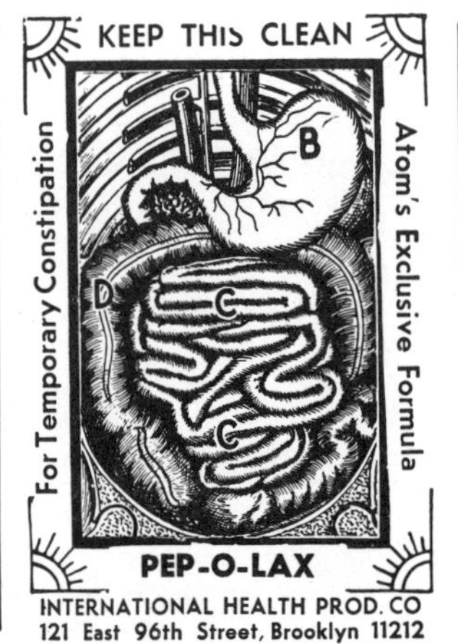

Atom's Exclusive Formula

PEP-O-LAX
INTERNATIONAL HEALTH PROD. CO
121 East 96th Street, Brooklyn 11212

DIRECTIONS
Adults: For temporary constipation one-half to one teaspoonful in half glass of water before retiring when necessary.
Children: 12 years or over ½ Adult dose. Contents: 1 oz.

tan Soap Company made large blocks of the product to his specifications. He would save expense by doing no advertising. Having had his own labels printed with Leah's picture on the wrapper, he then cut and packaged it himself. With his lack of overhead, he could sell a product of superb quality at four bars for only a quarter, and better could not be had for any price. In a couple of weeks, he was in a Broadway storefront selling "The Mighty Atom's Gold Piece Soap" by the case.

He interspersed feats of strength with his lectures. Anyone who considered his theories on strength and health to be mere verbiage could watch him do much of what he had performed in vaude. An average day would find him performing five to twenty-five numbers; one per lecture or an entire show of strength, depending upon his mood.

Standing shoulder to shoulder at any given time, one found college professors, gangsters, medical doctors (his biggest fans and most outspoken opponents—he referred to them as "MD's . . . mule drivers"), housewives, students, and children. Children especially loved to watch him. He catered to them, he understood them; in a curious way the Mighty Atom had remained a child himself.

"What is the hardest thing to get when invited into someone's home?" he asked them.

"Money?" a kid piped up from the front row.

The Atom shook his head. "People always ask you the same thing:

" 'Like a drink?'

" 'Fine. A glass of water, please.'

" 'How about a boilermaker?'

" 'A glass of water, please.'

" 'A cup of coffee?'

" 'Just water.'

" 'A Coke? Beer?'

" 'Water.'

" 'Champagne?'

" 'Water.'

" 'Cream soda?'

" 'Water.'

" 'How about . . .'

"People will offer you everything else in their home before they offer you a glass of water: the most readily available commodity that is virtually free, the one most healthful thing you should drink."

The children stood in the front row with mouths agape as he handed out steel bars bent into bracelets for souvenirs.

"Kids, I can see that some of you are standing before me with your noses running because you have colds. When you have a cold you don't have much fun, do you? I want you all to be at your best, so that's why you've often heard me ask you to quit eating 'cheap' sugar and junk food. You can get all the real sugar you need, the right kind, from raw fruit and vegetables. Now, we're going to see the proof of why you must eat right.

"Even though the snow is on the ground, and it's freezing outside, you may notice that when I leave here today, I will go home without an overcoat. I'm not showing off. I don't wear one because I don't own one. My wife and children have bought me several, but they just gathered dust in the closet until they were given away.

"And still, I don't have a cold . . . but you do. Why is that? Well, the reason is *our colds are caused by what we eat . . . caused by mucus-producing foods*. If you ingest these substances, your body will accumulate mucus until it must be discharged. That's really what a cold is, a cleansing process. That runny nose is more the result of what you've been eating than anything else.

"Sugar and bleached-flour products are mucus-producing substances. The person who eats processed sugar and white-flour products is preparing a nest of mucus that will attract germs, especially those which attack the upper respiratory system. During the most frozen winter, colds can be avoided by avoiding these mucus-producing substances like sugar and products containing it: candy, jams, jellies, chocolate, ice cream, soda pop; white flour and products containing it: white bread, cake, cookies, pies. Also, one should try not to eat fried or greasy foods.

"In short, avoid the typical American diet. Eat it, and you will catch colds with certainty.

"Don't acquire a taste for that which is not health-giving. I abstain from junk food and never catch cold. Never. I haven't had one since 1918. You can sneeze cold germs in my face, and I will not contract a sickness from it as would ninety-nine and nine-tenths per cent of the population, because my respiratory organs are free from the mucus which forms a breeding ground.

"So, you see, kids, I really don't need that overcoat because *colds are not caught . . . they are developed.*

"Now whichever kid comes back to me next week and has not eaten any junk food, I will give him or her a prize."

The Mighty Atom's words had a tremendous impact on young minds. He never turned away those wayward children who had been brought to him by distraught parents; he could accomplish what no peer, parent, or teacher could. Once the Atom took any youths in hand he showed them he liked and trusted them, he cajoled and threatened, he condemned, he hugged, he fed them and took them home, he made them work and paid them, he asked them for self-sacrifices and small personal favors, he pressed them into service as his assistants and made them stand tall in public, he gave them a sense of their own uniqueness and importance. Those he reformed in his lifetime numbered in the hundreds, white and black, Jew and Gentile, male and female. His most notable results were with hardened juvenile delinquents.

He kept a handbill posted in plain sight for parents to read:

THE MIGHTY ATOM'S TWELVE RULES FOR
RAISING DELINQUENT CHILDREN

How To Ruin Your Children—Guaranteed 100% Infallible

1. Begin in infancy to give the child everything he wants. In this way he will grow up to believe that the world owes him a living.
2. When he picks up bad words, laugh at him. This will make him think he is cute. It will also encourage him to pick up cuter phrases that will later blow off the top of your head.
3. Never give him any spiritual training. Wait until he is 21, and then let him decide for himself.
4. Avoid use of the word "wrong." It may develop into a guilt complex. Later, when he is arrested for stealing a car, he can conclude that society is against him and that he is being persecuted.

5. Pick up everything he leaves lying around; books, shoes, clothes. Do everything for him, so that he will be experienced in throwing all responsibility upon others.
6. Let him read any printed matter that he can get his hands on. Be careful that the silverware and drinking glasses are sterilized, but let his mind feast on garbage.
7. Quarrel frequently in the presence of your children. In this way they will not be too shocked when their home is broken up later.
8. Give a child all the spending money he wants. Never let him earn his own. Why should he have things as tough as you had them?
9. Satisfy his every craving for food, drink, and comfort. See that every sensual desire is gratified. Denial may lead to harmful frustration.
10. Take his part against neighbors, teachers, policemen. They are all prejudiced against your child.
11. When he gets into real trouble, apologize to yourself by saying: "I never could do anything with him."
12. Prepare for a life of grief. You will be likely to have it.

By this time, the word "pitchman" could hardly be applied to Joe Greenstein. The object in the pitch business was to speak for fifteen or twenty minutes, sell hard, and repeat the pitch and climax of sale as often as possible. In contrast, the Atom would speak for two or more hours at a time, with most of his talk having nothing to do with his products. Each long discourse cost money in time not spent selling, and other pitchmen thought him foolish.

Many of those in his audience were working people who frequently complained to him of muscle and joint pain, and rheumatic ailments. Having bought preparations for relief, they found nothing satisfactory. These complaints prompted him to create the next in his line of products, a liniment. Back to the books. Several months later, he had developed "The Atom's Superior Liniment." This was no pitchman's snake oil, but his registered formula of: crystals of menthol, artificial oil of mustard, oil of camphor, Eucalyptus, oil of cedar, kerosene, wintergreen, turpentine oil, linseed oil, Origanum. Intended for temporary relief of aches and pains, it proved a wonder for the longshoremen, housewives, weekend ballplayers, the cross-section of those who were his audience. The Atom's Superior Liniment would become one of his most popular products; his best customers were given a special discount when instead of buying the four-ounce bottles they ordered his liniment in gallon jugs.

The Mighty Atom continued to perform his Samson numbers, but now he offered his Atom's Ramsoil to the public, the identical hair-and-scalp formula he had developed for his own use in the days when he held back airplanes. After his lengthy discourse, he would briefly demonstrate or describe his products. Knowing that they were of superior quality and fairly priced, he indulged in no manipulation. After explaining what they could and could not do, he offered them once. If you bought . . . fine; if not . . . fine. If after listening, you chose not to buy, that was your own business.

One of the foundations of the Mighty Atom's philosophy of health and vigor was that incorrect diet and the resultant faulty elimination were responsible for much of human disease. Theories which would gain widespread popularity in the 1970s were offered by Joe Greenstein as early as a half century before.

"Anyone can wash with soap and water, and think they're clean. Nonsense. The only real cleanliness begins with INTERNAL CLEANLINESS. And when you practice that, the rest is easy. When you eat right, you feel right, you live right, you think right. You enjoy life in the right way. And this internal cleanliness aids the chain reaction of mental, spiritual, as well as true physical cleanliness. *A human being should empty his bowels at least once a day.* Cleansing of the bowel is a foremost preventative of sickness, and has ramifications in every area of human activity.

"Just as the gizzard of a fowl needs the mechanical action of sand and gravel mixed with food, so does the human digestive system need roughage on a daily basis to help the intestines cleanse themselves. Lack of roughage prevents normal muscular expansion and contraction of the intestines which we call peristaltic action.

"Constipation is an ailment that most people suffer from. It is a result of improper diet. Most people don't even know they're constipated, because they've been that way for their entire lives.

"No food should remain in the body longer than twenty-four

hours. Yet, some people eat three meals a day and move their bowels once a week. And they wonder why they feel bad. During the day, the constipated individual lacks energy, feels sluggish, heavy. He doesn't know it is the result of what he has and has not eaten.

"The awful odor of human excrement is a result of putrefaction, of its being retained too long before being discharged. Such would not be the case with a proper cycle of elimination. The unpleasant smell of human sweat and human waste is merely an indication of what is going on inside.

"Once that which was food leaves the small intestine and goes into the ascending colon, it is no longer food, but waste. If it is not passed along correctly, but is kept in the body too long, toxins will accumulate in the intestine before finding their way into the bloodstream. Constipation is 'self-poisoning.' It will cause aches and pains, headaches, and promote conditions of ill health.

"But what do we do? We refine and process our foods, so that the very thing we require is discarded.

"How does most white flour become white? Golden wheat is harvested, ground, and the bran content taken away in the manufacturing process. That all-important husk is thrown to cattle and farm animals, who incidentally are never constipated.

"Our staple of plain white flour when mixed with water becomes a cheap paste. In the intestines, about the same thing occurs. The resultant pressure creates hemorrhoids, and if allowed as a lifetime pattern can create consequences more dire. Therefore, in my opinion, such foods as white bread, cake, cookies, and the like are junk. Americans are not constipated, they are 'consti-pasted.' Clogged up. The stuff isn't called 'paste-ry' for nothing.

"Instead of foods that are polished and cosmeticized, I advocate the eating of raw fruits and vegetables, whole-grain natural foods, celery, bran, wheat germ, whole-grain breads, a fresh apple before bedtime. Man cannot improve upon the original article. To the degree that man has processed these foods, to that degree should they be avoided. Eating of these natural foods with the substance left intact will help prevent conditions ranging from faulty elimination to colonic cancer.

"When a dog feels bad, he will go to the garden and eat grass. This is a purgative which nature tells him is not harmful. The average human unfortunately doesn't have the sense of a dog. When we are feeling poorly or constipated, we go to a drugstore and eat a harsh chemical. To my way of thinking, common chemical laxatives are harmful to the human system; they irritate the lining of the bowel. A diarrhetic movement is created by drawing fluid from the blood. After the resultant unnatural elimination, one stumbles from the bathroom with a feeling of weakness."

The Atom saw that those who would not heed his dietary suggestions needed a good, safe, natural laxative that was not a drug. He had for many years studied the properties of herbs, roots, barks, berries, seeds, leaves, blossoms, gums, vegetables, fruits. Now he spent evenings at home and in the library pouring over books, devising an herb tea. In his typical style, his daytime lectures advocated a diet which, if followed, would do away with the need for the very product he was creating. He gave handbills to his audience:

THE MIGHTY ATOM'S
SIMPLE RULES FOR SOUND HEALTH AND LONGEVITY

* Moderation. Do not overload your stomach. It is not a garbage can.
* Forget the clock. Eat one or a half dozen small meals a day . . . but only when you are hungry.
* Fast at least one day a week. Rest your bodily machinery and double its life.
* Drink pure water in sufficient amounts.
* AVOID:
  Constipating, mucus-producing foods, sugar and bleached flour products, greasy and junk foods;
  The heart-killers: fatty foods, consumption of red meats and egg yolks to excess;
  Fats, sugar, and bleached-flour products, salt, coffee, tea, preservatives, and denatured foods.
* EAT:
  A generous daily diet of raw fruits and vegetables which have been well washed:

Nonfat dairy products, fish, poultry; any red meat should be lean;
A daily amount of roughage for sound digestion: raw fruits and stalk vegetables, whole grains, bran.

* Move your bowels at least once a day.
* Use a good vitamin supplement.
* Daily exercise after a warm-up. Slow, steady progress.

If this regimen seems unpleasant to you, do not adopt an all-or-nothing attitude. *If you can't do the best thing for yourself . . . do the next-best.* Implementing any one or all of the above will pay big dividends.

After about six months, he had worked out the basic formula for his herb tea. It would take two and a half years to perfect. In the tradition of the selfless scientist, he experimented upon the ultimate guinea pig—himself. The prescribed dose, a teaspoonful in a half glass of water, frequently resulted in a sprint to the commode. But finally he was satisfied that the product would work safely and comfortably. He gave it the brand name "Pep-O-Lax."

On one side of the box was a diagram of the human digestive tract, along with the admonition, "Keep This Clean." The thirteen ingredients were plainly marked on the package:

Sassafras Root	Althea Root
Hungarian Chamomile Flower	Alfalfa
Caraway Seed	Berberis Acquit
T. V. Senna	Licorice
Chicory Root	Coriander Seed
Peppermint Leaves	Spanish Aniseed
Golden Seed Root	

Every box of his "herbs" carried a photo of the Mighty Atom. He would dispense Pep-O-Lax herb tea and his other products, in person and by mail order, to a faithful following of thousands for the next fifty years.

So Joe Greenstein became the King of the Pitchmen. Nowhere else was there a man who could lecture and perform in a furnace of

August heat for twelve hours, without once stepping down. Despite the Depression, his days of starving were over. He made a habit of rounding up his out-of-work vaudeville friends a dozen at a time and taking them out for a square meal. He continued with his charity shows, and would lend his services to any group that had need of him.

Winters he occupied storefronts in Manhattan, Brooklyn, or The Bronx. On Broadway, his marathon lectures did not go unnoticed by the press. He appeared frequently in the columns of Mark Hellinger, Walter Winchell, and others; Winchell referred to him as "the best show on Broadway."

Summers, he would pack Leah, the kids, and a half dozen trunks into a huge black Packard, and off they would go to Saratoga Springs. The protesting car sagged to the bottom of its suspension, what with the children and his paraphernalia including two 250-pound barbells strapped from each running board up across the fenders.

In Saratoga Springs, he would spend the summer lecturing and selling his products at Putnam and Phila streets, ensconced in one of the many spa hotels, or exhibiting at Convention Hall.

He divided his time then, working the fairs and farmers markets of Pennsylvania Dutch country which he found particularly to his liking for its beauty, charm, and the civility of its people. He had been a regular visitor there since 1927.

Now as he established a routine, each weekday found him in little towns like Perkiomenville, East Petersburg, Leesport, Schuylkill Haven, Boots Corner, Gilbertsville. His reputation began to grow at the large rural markets like Zern's and the Green Dragon. As the thirties began, he spent much of his time in those rural Pennsylvania fairs, but also found himself drawn closer to home, to the great playground . . . Coney Island.

Coney Island in those days was a miraculous place, a land of wonders like Steeplechase Park, where brightly painted wooden horses raced on rails around and through the pavilion with gleeful children upon their backs. There had been Dreamland, a Beaux-Arts fantasy in stucco, by whose entrance stood a pair of dragons, their wings alight at night.

But of all the Coney wonderlands, by far the greatest was Luna Park. Dignified, sophisticated, holding all manner of charm and romance, this City of Lights was singular in all the world. Russian novelist Maxim Gorki wrote of it at night:

> Thousands of ruddy sparks glimmer in the darkness, limning in fine sensitive outline on the black background of the sky shapely towers of miraculous castles, palaces and temples. . . . Fabulous beyond conceiving, ineffably beautiful, is this fiery scintillation.

It was to this Coney Island that Greenstein came, adding himself to its wonders.

He opened a place of business at Boardwalk and Twenty-third Street, next to Nedick's and Perry's Cycledrome, where highly powered motorcycles raced around the walls of a huge wooden cylinder. A perfect opportunity to publicize his grand opening presented itself when he discovered a two-story house being relocated. The house had already been lowered onto a wheeled platform, and the Atom towed it, with his hair, one full block down Coney Island Avenue.

After renting a small apartment at Twenty-eighth Street and Boardwalk as a family convenience, he gilded his establishment with a display of several hundred magazine and newspaper articles, action photos, Mighty Atom vaudeville posters, medals and keys to twenty cities, a startling exhibit of mangled, bitten, twisted, and broken iron bars, chains, and horseshoes, notarized letters from a score of dentists attesting to his seemingly impossible ability to bite through iron, the bullet which Dr. Daly had removed from between his eyes, presented along with the press report of the incident. The lecture stand was the repository of his life's artifacts.

Joe nailed up his sign:

THE MIGHTY ATOM
WORLD'S STRONGEST MAN
Lectures—Demonstrations

A few feet away on the boardwalk a large man sourly observed Joe setting up. His well-over two hundred pounds were stuffed into a

Warren Lincoln Travis, wearing the Richard K. Fox Diamond Belt for World's Heavyweight Lifting Championship

three-piece suit. Despite his huge stature and apparent anger, he had a pleasant, almost babyish face.

"Hey," he called to Joe. The Atom turned and the big man affixed him with a deadly stare. "Don't unpack your bag. You're not gonna be here that long."

"Yeah? Who says so?"

"Me. Warren Lincoln Travis. The Strongest Man in the World."

This was another of the most remarkable strongmen of the century; Warren Lincoln Travis was the holder of the Richard K. Fox Diamond Belt for the World's Heavyweight Lifting Championship. He was recognized by the *National Police Gazette* as "The Strongest Man in the World." In his youth, at 5 foot 8 inches and 185 pounds, Travis could backlift the staggering weight of 4,200 pounds. He could lift 1,105 pounds with two fingers. These were only a few of his world records. As a performer, he had executed such feats as supporting a carousel with fourteen men on it, allowing an automobile to run over him, holding a cannon upon his shoulders while it was fired, backlifting one thousand pounds a thousand times in quick succession, and other stunts equally fantastic. The bars of Travis's show weights were of such girth that only he, with his thick hands and superhuman grip, could lift them.

As a young man in Galveston, Joe had read about and corresponded with this man who was then internationally recognized, having established world records as far back as 1907. The two were friends though they had never met in person.

"You know," Joe said, "I wasn't always known as the Mighty Atom. My real name is Joseph L. Greenstein."

"Joe?" For a moment Travis forgot his anger, and offered his hand. "Why didn't you say something? I didn't know it was you. But listen, you have to take that sign down." Travis was adamant. He so zealously guarded his title that he maintained a standing offer of ten thousand dollars for anyone who could duplicate his feats. Now Greenstein had dared to display the coveted title only a few blocks from where Travis himself performed. "Joe"—Travis took a handbill out of his pocket and held it up—"can you do any of these exhibitions?"

"No," Joe readily admitted, "I can't do one of them."

"Well, then . . ."

"Warren"—the Atom pointed to one of his own posted vaude playbills—"can you do any of these?"

"No, I guess not," Travis said reluctantly after thoughtful perusal, though still in no way abandoning his claim.

Both men were right. The question of who is the strongest man in the world can best be answered with another question: strongest at what? The title may be applied to that individual who can lift the most weight. But there are innumerable lifts and countless other tests of strength that could be grounds for such a claim. In fact, there are many men who are "The Strongest Man in the World" at a specific feat of strength. No one man can do them all. The Atom was unequaled at bending, breaking, biting, twisting, and driving. Travis was a marvel at specific feats of weight lifting, particularly back and harness lifts. In truth, both were the strongest men in the world at their particular endeavors.

"Listen, Warren," Joe suggested, "we're gentlemen and friends. I can tell you that I've had all the glory I'll ever need. I'm not here to challenge you, only to lecture and make my living. I won't force the issue, if you won't."

"Okay, Joe," Travis agreed, "we'll leave it at that."

It was a pleasant enough summer day at Coney, though the July heat baked the rest of Brooklyn. At his lecture stand a strong, warm sea breeze blew in from the water. Business had been good that weekend afternoon, yet the pace of the day had been slow and uneventful. It was July 14, 1932.

He was alone, preparing to give another lecture. He began gathering a crowd by performing a few perfunctory feats, the people stopping with interest as he warmed up with repetitive one-hand lifts of his shot-filled barbell. He stopped suddenly, the weight poised mid-air. He could feel heat through the soles of his shoes. He looked down; through the cracks in the floorboards he could see the fire whipping in from the understructure of boardwalk beneath the building. His small audience stared blankly at him.

"Run," he said, then tossed the barbell down as the first flames burst through the floorboards, smoke pouring up around him. Rocked backwards like a prizefighter, he retreated and jumped through the open window at the rear, landing on hands and knees in the alley ten feet below. At the foundation of the boardwalk he could see great sparks and tongues of orange-blue flame. He had to go back. A whole life's accomplishment was in that lecture stand, its contents irreplaceable. He darted around Nedick's and up onto the flaming walkway. People were screaming, running aimlessly.

Smoke and fire were already belching from his lecture stand, the floor and walls engulfed in a swirl of orange. He fought to get in, but a wall of heat drove him back before the flames touched him. The forty-mile-an-hour sea breeze had whipped the tinderbox of boardwalk and stalls into an inferno. In a few minutes, three entire square blocks north to Surf Avenue were a howling blaze.

Everything of sentiment, everything of livelihood, all his memorabilia, all his wares: lost. It was worse than Newtown. So intense had been the heat that even the weights and metal artifacts had seemingly been vaporized. Not even a cinder was left. An hour later, there would be absolutely nothing to indicate that anyone had ever occupied the site.

The New York Times of July 14, 1932, reported:

> $5,000,000 CONEY ISLAND FIRE
>
> . . . 1,000 homeless . . . in the three blocks between 21st to 24th Streets from Boardwalk north to Surf Avenue, nothing remained standing. . . .

Joe Greenstein hated the taste of whiskey. But now he bought a bottle and for the first time in his life got stinking, falling-down drunk. Warren Travis found him in one of the emergency shelters lashed to a cot. The volunteers there had been forced to tie him down as he raved senselessly before passing out. If ever a man had an excuse for getting drunk, the Mighty Atom did; the Great Coney Island Fire of 1932 had started right beneath his feet.

Some weeks later, he had scraped up enough money to open another place on the Boardwalk near the Half Moon Hotel. Though his artifacts and treasured mementos were gone, he ordered more Mighty Atom products and began again.

It was on an afternoon that August that the Atom took a break from his lecturing and strolled down the boardwalk and over to Coney's Bowery to see his friend Travis. There Joe presented him with a hefty piece of rubber. The Great Travis had always to be on the lookout, as Greenstein was constantly coming up with new and bizarre tests of power.

"What's this?" Travis inquired.

"It's just a stretch exerciser. I made it." Indeed, Greenstein had made it himself, a monstrous rubber band of inner tubes laminated one upon the other. The chunk of rubber was 12 inches long and ½ inch thick.

Travis set himself, took either end in his enormous grip, and strained to stretch it across his chest. He could hardly budge it; it was like trying to elongate a truck tire.

When Travis had had enough, the Atom stretched the chunk of rubber out to his full arm span of over 6 feet. Satisfied that he had not been tricked, Travis commented with respect, "Atom, there's not another man alive who could do that. I know."

Forty-five years later, the Mighty Atom would still cherish his homemade "rubber band" as a memento of his association with the Great Travis.

In the cool night breeze, with the neon of the amusements behind them, the two strongmen stood at the boardwalk railing looking out at the moonlit surf. Travis made a frank admission.

"Atom, I've seen a lot of men in my time, and beaten the lot. But you . . ." The giant's face revealed a certain vulnerability. "You're a mystifying little man."

"I'm not a little man, Warren. There's no such thing. No man has limits, except in his own mind. Look up, what do you see?"

"Stars . . . ," said Travis.

"Stars, planets, whirling infinity," the Atom continued, "with no beginning and no end. Enough to make a man lose his mind. My deity was never born and never died. He Was, Is, Will Be. My God is Space and Time."

"Mystifying . . ." Travis shook his head.

Warren Lincoln Travis was a very private man whose real name was Roland Morgan. He was resolute and self-made with a stoic exterior; but beneath it was a gregarious person, somewhat sentimental, with a keen mind. Like many who had weathered a difficult childhood, Travis was known to be a close man with a dollar. Yet Greenstein's experience with him was just the opposite. A loner, he was attracted by the warmth of the Greenstein home. With great affection, he often invited Joe, Leah, and all the kids out for a lavish dinner.

"Does everybody have what they want?" He made sure the children were served before he dug into his own double blueplate special. "One thing about a stray dog or an orphan boy . . . ," he said philosophically, "he always knows when he's well off."

"Were you an orphan, Warren?" the Atom asked.

"Yes, I was," he said quietly. When the meal had concluded, the big man reached into his breast pocket, produced a fat billfold secured by a rubberband, quickly riffled through a small fortune in bills, and paid for the meal.

"Don't you believe in banks?" Joe inquired.

"Banks fail. People rob banks." Travis returned the billfold to his breast pocket. "Everything I have, that and my title I carry right here." He patted the money over his heart. "Anybody thinks they can take either . . . let 'em try."

Travis never married, and eventually amassed a sizable fortune through real estate investments; but because he didn't trust banks, he converted his cash and assets to the "hardest" currency: diamonds. As he advanced in years, Joe and Leah implored Travis to take things easy. But he would not. With no wife, no family, his life was his title, reputation, and his ability to perform. Though well into his sixties, he continued with endurance feats such as his famous one thousand pounds one thousand times, quickly raising the weight as the throng chanted out the numbers.

On July 12, 1941, he was performing as usual at the World's Sideshow, 1216 Surf Avenue, before a typical Saturday-night Coney Island crowd. Across the street was Luna Park, as always its softly lighted parapets appearing like fireflies against the night sky. The air was filled with calliope and laughing sounds. The cotton-candy man spun his sweet threads as the crowd of revelers wandered through the gay kiosks—sailors and their girls, old men and grandchildren—with prizes of carnival glass.

Travis knelt and one-handed one of his huge chrome show weights overhead, standing like Atlas with the world upraised. With the crowd in silent awe, he held it aloft, the lights of Luna Park dancing off the gleaming steel globes.

"You know, Atom," he had said only a week before, "when I was a puny little kid, I always wanted to do something that nobody else could do." His face was etched with the burden of a half century at his formidable profession. "I've held a dozen world records. A dozen. I guess that's enough."

He repressed a wince of pain, and endured a long moment, but at last, inevitably, the weight and the man descended with a crash. Arm still outstretched, like a felled tree, the Great Travis was no more. In the background could be heard the cackle of the mechanical funhouse clown, the tinny strains of the calliope. The Strongest Man in the World died onstage at sixty-five, performing the same feats he had at the turn of the century.

The strongman who didn't trust banks had hidden his fortune well. To this day, the diamonds of Warren Lincoln Travis have never been found.

NAZI BASEBALL

The Atom held a curious attitude toward his audience. Unlike H. L. Mencken, whose dictum "No one ever went broke underestimating the taste of the American public" revealed an obvious disdain for the common man, Greenstein maintained a curious optimism.

"I never knew if the man standing before me was a Supreme Court Justice or a convicted felon, a man of intellect or an imbecile." Having had them all in his audience, he preferred to give everyone the benefit of the doubt. As his message was one of self-improvement, he would not look askance at the most unlikely potential recruit.

But in every random group there was that small cross-section of loudmouths, drunks, show-offs, anti-Semites, and assorted rabble; and in dealing with these, his position was clear. "I give you respect. But I do not ask for yours in return. I demand it. If you will not give it to me, I will take it out of your hide." He would go to great lengths to protect the sanctity of his little world and would not permit "hoodlums" to interrupt his thoughts, or spoil his audience's education. He attempted in all ways to maintain peace and tranquility, but if he were forced to step down from the platform, one of his firm hands atop a man's shoulder was usually sufficient to give pause. However, any anti-Semitic epithet was guaranteed to put an immediate end to his restraint; and whether the insult to his people came from one man or a dozen, the reply would be the same—immediate retaliation. Al Spielman, the author's father, recalls watching the Atom humble a vociferous anti-Semite, extracting a public apology from the man with a wrist joint lock, administered gingerly with two fingers, as if handling a butterfly.

"I am a peaceable but by no means passive man," the Atom would say. "I consider mine to be important work. And I am not the kind who will accept abuse."

During Coney's heyday Sam Spielman, the author's uncle, owned a home there on West Twenty-eighth Street and, as a frequent visitor to Prof. Greenstein's lectures, had the opportunity to witness one of his fabled one-round bouts.

It was summer, and as usual the crowds came on this particular Sunday afternoon to see the great strongman. In the midst of detailing one of life's more obscure health secrets, he was repeatedly interrupted by a trio of hulking toughs, each with regulation tattoo and tank-top T shirt. They taunted and heckled with loud insults. After repeatedly requesting them to be quiet, which only incited them, he told them to "take a walk." They threatened him. At last, they committed the unpardonable, shouting obscenities amid an audience largely composed of women and children. Today Sam Spielman, a spry, lanky man in his sixties, recalls the incident with vague disbelief.

"These were three barroom-brawler types each almost twice his size. As I blinked, the Atom, already off his platform, threw three punches. The three wise-guys were out stone cold on the boardwalk." The cops had to dispatch an ambulance to haul them away.

So numerous were such brawls that the little Samson could not even recall this particular incident. "I had to do that a couple of times a day for decades in order to earn my living," he said. "I should remember *those* three hoodlums?"

One he would recall for its amusement value was the time he knocked out two "nogoodniks" and deposited them on the scale of the "guess your weight" man in the next booth. He guessed, and missed the weight by seven pounds. Or the time that a bunch of drunken conventioneers had showered his gathering with debris from their window in the Half Moon Hotel. (The Half Moon had gained notoriety as the last perch of Abe "Kid Twist" Reles; a gangster turned informant, Reles was under police guard when he fell or was pushed from a high window, and was popularly referred to thereafter as "the bird who could sing, but couldn't fly.") When at last a half-grapefruit skin pelted Joe's audience, he stopped the lecture, climbed like Tarzan from the roof of his own stand to the hotel, and dove through the window where the offending party was quartered. He chased them out of the Half Moon, and down the beach.

A young man named Vic Boff, who would later make his own mark in athletic and health circles, recalled one of the Atom's visits to New Jersey.

"I had come to Newark to hear Joe lecture. There was a 'Neanderthal man' in the crowd, maybe six foot four, very crude and boisterous, mimicking Joe in a most derogatory way. He kept this up for about fifteen minutes after repeated requests for quiet. Finally, the Atom could stand no more. He charged off the platform, the quickness of his attack in itself a blur, hit the big man one shot on the button, spun him around, and deposited him neatly in the gutter. Case closed . . . for the moment.

"At the end of his lecture, the Atom was seated in a chair for a demonstration wrapped to his neck in chains. Now the big man suddenly reappeared with a perfect opportunity to get even. The little Samson was helpless; Neanderthal made for the platform.

"Joe saw him coming and let out with a piercing, blood-curdling bellow that made everyone's hair stand on end; and the chains started flying off. The whole scene was so wild, so demonic, that the big man stopped in his tracks . . . and then ran away."

Should anyone doubt these accounts, no less than seven New York newspapers noted one of his more famous encounters. When not lecturing, Joe Greenstein sometimes took a busman's holiday and enjoyed listening to other health practitioners. He and son Harry went to the Health Lecture Auditorium, at 841 Eighth Avenue. They tried to listen attentively, but the crowd was very rowdy. Three men in particular kept up a stream of abuse. Finally, Harry stood and told them to shut up.

Harry Greenstein had in his teens been severely crippled in one leg as a result of a traffic accident, but he was by no means the Tiny Tim of the family. At twenty-three, he was now a strapping and muscular young man, the largest of the Greenstein boys, and though a gentle and easygoing sort, he was his father's son and was not known to back down from a confrontation.

The three men strode over to Harry and the fortyish little bearded man who now suddenly rose beside him. When, after a few pleasantries, one of them punched Harry in the mouth, the 145-pound strongman went into action. What follows was recorded by the New York newspapers on March 3, 1937:

> The Mighty Atom's first swing put Osmond Roach, twenty-seven years old . . . of 109 West 118th Street, out for half an hour. Frank Doliva, twenty-three, 1505 York Avenue, was next. The third man fled. Meanwhile other spectators joined the fracas and threw Harry through a plate glass window.
>
> —New York *Post*

> MIGHTY ATOM PROVES POWER IN FREE FOR ALL
>
> —NEW YORK *Journal*

> ATOM FLATTENS FOES, LITTLE STRONG MAN TACKLES CROWD OF HECKLERS
>
> —The *Sun*

The newspapers reported that Harry had been stabbed, and that his father had then taken on the entire crowd. What they had not reported were the details. Greenstein had, in the space of a few minutes, tossed several combatants through the plate-glass window through which Harry had made his exit; knocked out a score of others, sending several to the hospital; and when the fight spilled out into the street, hurled the wielder of the knife onto a moving taxicab. There is no disparity between reports of his having struck

The Sun

NEW YORK, WEDNESDAY, MARCH 3, 1937.

ATOM FLATTENS FOES

Little Strong Man Tackles Crowd of Hecklers.

New York American

WEDNESDAY, MARCH 3, 1937

'MIGHTY ATOM' THRASHES TWO

NEW YORK JOURNAL

WEDNESDAY, MARCH 3, 1937

'Mighty Atom' Proves Power in Free-for-All

M. A. GETS ANGRY

ANGER GROWS.

דער מארגען זשורנאל

MARCH 5, 1937

New York Post

WEDNESDAY, MARCH 3, 1937

MIGHTY ATOM KNOCKS OUT TWO LECTURE HECKLERS

New York World-Telegram

NEW YORK, WEDNESDAY, MARCH 3, 1937.

Police Had to Rescue Crowd from 'Mighty Atom' in Fight

few blows but having rendered great damage. Whether it was arm or head, whatever the little giant hit would break.

The two original assailants were booked for felonious assault: Harry got patched up, and he and his old man went home. The Atom would never take any joy or glory in such an incident, but one New York paper had said it best:

> POLICE HAD TO RESCUE CROWD FROM "MIGHTY ATOM" IN FIGHT
>
> . . . When the police arrived Mr. Greenstein had the situation well in hand. As one cop put it, "He didn't need no help. We had to rescue the crowd from him."
>
> —New York *World Telegram*

At this time the Atom was asked to appear on a national radio broadcast to be sponsored by one of the nation's largest meat-packing firms. He accepted the offer, rehearsing enthusiastically in preparation for the event, for which he was to be paid a substantial sum.

He arrived at the studio with an entourage of his three oldest sons and several trunks of strongman gear. The "ON THE AIR" light was already on. Someone slapped a copy of the format and last-minute script changes into his hand. He listened offstage as he was introduced.

"The Mighty Atom . . . ," the MC declared, "has become one of the strongest men in the world . . . *by eating raw meat!*"

The near-vegetarian glanced down at the script, which indicated that he was to attest to this canard at the end of his performance.

"It's a lie! . . . a menace to the public!" he railed backstage. Everyone there tried to hush him up, lest his comments be picked up and broadcast. The producer raced over to placate him.

"Please, Mr. Atom . . . I'm sure you're overreacting. . . ."

"You know what I'm going to say, if I ever get on. I'm going to say what I think: *If you love animals, don't eat them!* and I don't give a damn who the sponsor is."

"My God"—the producer went white—"that's out of the question."

"Harry!" the Atom snorted to his oldest son, ". . . pack up, we're leaving!"

The producer was confronted with the terrifying prospect of five minutes of dead air. "Mr. Atom, you can't leave—you're going on in less than one minute."

"*No, I'm not. I won't be a party to a fraud. You'll have to make a retraction about that raw-meat business. . . .*" Hardly what one would expect of the sponsor. "The truth is . . . ," he expostulated, "*too much meat . . . will kill you. It will choke your heart with death-dealing fat. That's what I'm going to say!*" The Mighty Atom and sons roared out of the studio, leaving behind the engagement and the money.

If radio was not his medium, there were still sporadic theater bookings. While playing a successful one-week limited engagement at Philadelphia's Arch Street Theatre in January of 1935, for his concluding performance the Mighty Atom accomplished one of the most stunning feats of his career. Unlike others of his repertoire, he would perform this only once.

His concluding performance at the Arch Street was standing room only in anticipation of the announced event, which even for the Atom seemed impossible. The finale curtain opened, and there as promised was a solid steel I-beam 30 feet long, 6 and 8 inches wide, and almost ½ inch thick. It weighed about 700 pounds.

As the crowd hushed, twenty volunteers stepped onstage. He positioned ten men at each end of the beam, five on either side facing one another. Burlap sacks were thrown over his shoulders for protection. He took his firm stance with one foot forward, knee bent, the other leg back and locked. He set himself. At his signal, the twenty men hoisted the beam and positioned it across his neck and shoulders.

As the audience became deathly quiet, he shouted, "*GO!*" The men released the weight and pushed down. As the Atom stood firm, the audience watched, fascinated, as the steel slowly caved in, its ends levering down towards the floor. When both ends rested on the stage, he stepped out from beneath the weight and let it topple forward with a thunder that shook the theater.

The event was heralded as the most remarkable of feats, a banquet was tended in his honor by a group of local athletes, and an enterprising Philadelphia junkman obtained the mangled beam, displaying it for twenty-five cents a peek as if it were the Cardiff Giant.

The Atom was a bit embarrassed by the reaction. In truth, it had not been the most difficult of his feats. Far from it. Bending a short bar 9 inches long was infinitely more arduous. He had proven his ability to hold up substantial weight by raising 3,300 pounds on a backlift apparatus. This beam weighed only 700 pounds, and while it appeared gigantic, the steel was a malleable variety. He knew that if he could support it atop neck and shoulders, the rest of the weight would be suspended by and focused against that very center point. The beam would practically bend itself. With a little help on either side (twenty men weren't really required), the ends would be crushed to the floor.

In fact, he had not devised this feat. As a boy of fourteen in

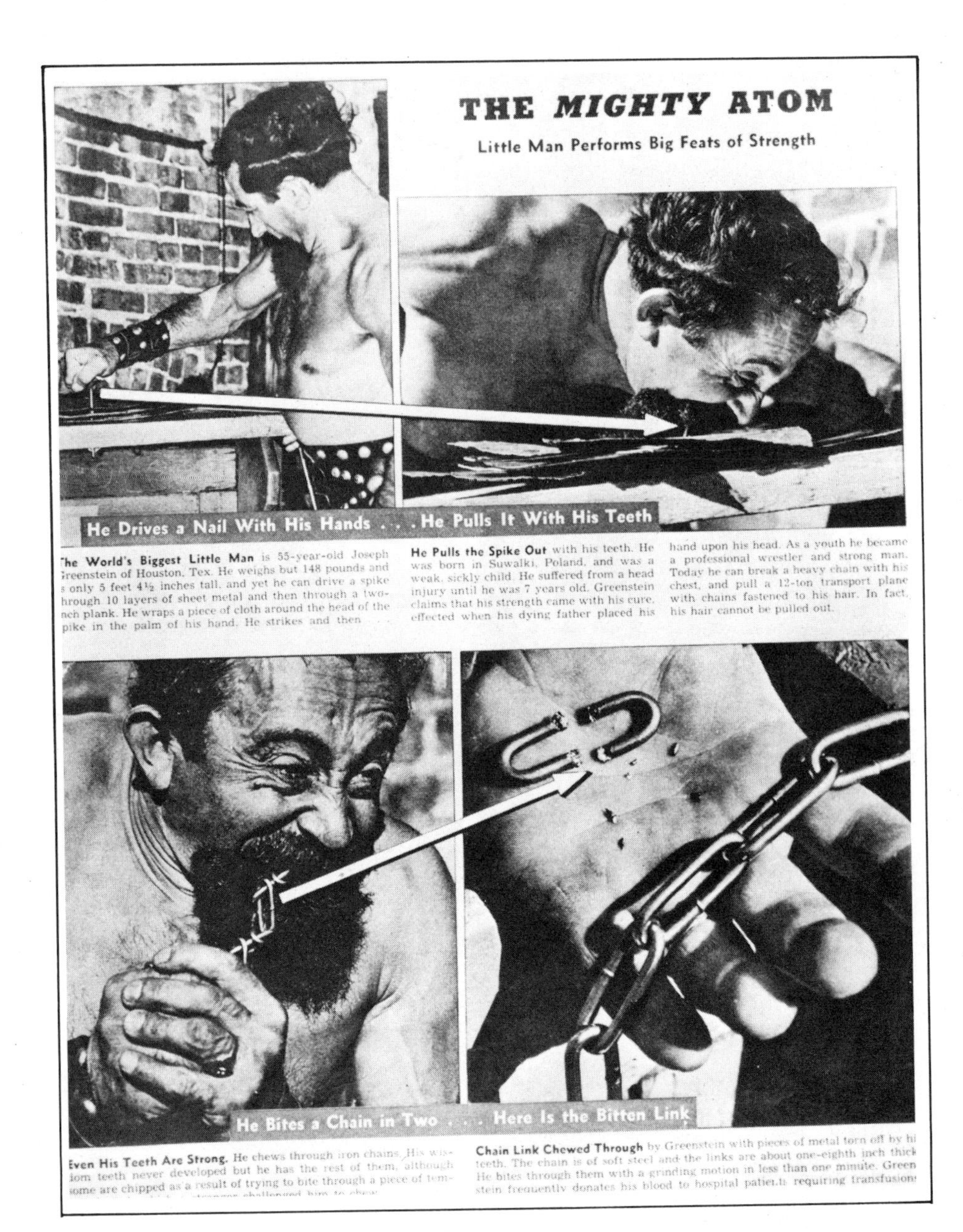

THE *MIGHTY* ATOM

Little Man Performs Big Feats of Strength

He Drives a Nail With His Hands . . . . He Pulls It With His Teeth

The World's Biggest Little Man is 55-year-old Joseph Greenstein of Houston, Tex. He weighs but 148 pounds and is only 5 feet 4½ inches tall, and yet he can drive a spike through 10 layers of sheet metal and then through a two-inch plank. He wraps a piece of cloth around the head of the spike in the palm of his hand. He strikes and then . . .

He Pulls the Spike Out with his teeth. He was born in Suwalki, Poland, and was a weak, sickly child. He suffered from a head injury until he was 7 years old. Greenstein claims that his strength came with his cure, effected when his dying father placed his hand upon his head. As a youth he became a professional wrestler and strong man. Today he can break a heavy chain with his chest, and pull a 12-ton transport plane with chains fastened to his hair. In fact, his hair cannot be pulled out.

He Bites a Chain in Two . . . . Here Is the Bitten Link

Even His Teeth Are Strong. He chews through iron chains. His wisdom teeth never developed but he has the rest of them, although some are chipped as a result of trying to bite through a piece of tem-

Chain Link Chewed Through by Greenstein with pieces of metal torn off by his teeth. The chain is of soft steel and the links are about one-eighth inch thick. He bites through them with a grinding motion in less than one minute. Greenstein frequently donates his blood to hospital patients requiring transfusions

Look magazine, June 1938. Courtesy *Look* magazine

Suvalk, Poland, he had seen a Jewish strongman by the name of Morrie Chehovsky perform a similar one. That Chehovsky was an athlete who approached 300 pounds did not concern him. If the Mighty Atom could simply support the weight he would succeed, and he did.

Having been out of the vaude limelight for some years, the Atom nevertheless continued to attract national attention.

In early 1939, with a towing record of thirty-two tons of trucks, he attempted to pull a *DC–3 airliner* from a standstill. Having failed to get it moving, he enlisted the aid of five aircraft personnel who pulled on his arms to get the silver bird to break its station. Instead, they nearly broke his neck. Later, examining the tires, he concluded that insufficient inflation had caused too much rolling resistance.

Undaunted, he decided to attack the tonnage of all time—the *Queen Mary*. A gray winter's day found him on a pier at Manhattan's Battery Park, surrounded by police and anxious photographers. He planned to stand on the tugboat's deck with a line from the *Queen Mary* affixed to his hair. The tug would move forward pulling the liner into port. Fortunately, the event never came off. Before Atom and entourage could pile into the tugboat, the feat was canceled, washed out by a torrent of freezing rain.

That summer, he was approached by *PIC* magazine to attempt a feat in which his life would hinge on his well-known ability to bite through and tear chains. He accepted the challenge contingent upon the sum, five hundred dollars, being paid in advance.

After receiving his money, the Atom found himself on a stretch of road in Brooklyn's then rural Bergen Beach section. In the middle of the roadway he seated himself in a chair and several assistants shackled him to it with iron chains around his head, arms, legs, torso, a loop of chain between his teeth, another encircling his deflated chest. The chains were of a test strength of 300 pounds.

An official stepped forward and impassively snapped several padlocks shut, securing the iron traces. "Atom, you've got a minute and a half. That's all," he said, and checked his starter's pistol.

A mile away, a man gripped the steering wheel of his auto, and raced the engine.

"*Ready* . . .," the official called to the chained Atom. Photographers raised their cameras. The official held his starter's pistol high and fired. The car jack-rabbited away and came on at forty miles per hour. The Atom would have ninety seconds to sever the chains, free himself, and jump clear before the car crushed him.

The magazine had come for action shots; the driver, fortified by a little "hair of the dog," was instructed to let the strongman earn his money. The driver would not use the brakes. If the worst happened, the magazine's circulation would simply go up.

As the photographers clicked off their pictures, second by second, the Mighty Atom strove to escape. The first chain was burst quickly by chest expansion; he bit through the second with some difficulty. He had expected to have time to spare. No such luck. In the heat, the sweat from his hands hampered his grip; but he quickly tore through the third and fourth chains as the big machine loomed up over the rise and headed down the last stretch. He ripped the last of the iron bonds from his legs as it was almost upon him. He could see the driver's face, a flash of grillwork, a hood ornament. . . .

He jumped clear as the car sailed in, the chair flying to pieces, the vehicle catching his foot mid-air and spinning him around. He

BROADWAY

3 MAGAZINES IN ONE • • COVERING THE ENTIRE FIELD OF ENTERTAINMENT

BY THE SKIN OF HIS TEETH

Joseph Greenstein, from Houston, Texas, has developed his own way of making news. Unlike Simon Hertzberg, the Danish carpenter who thinks nothing of letting a car run over him (see pages 36 and 37), Greenstein makes his bid for fame by not letting a car run over him. He sits shackled to a chair in the path of a car speeding toward him at 40 miles per hour. The chains, which bind his arms, legs, head and body to the chair, are of steel and have a tested strength of more than 300 pounds. They are knotted and twisted so that he must break them before he can escape, and all within a time limit of less than two minutes.

Seated in the chair he faces the approaching car. One loop of the chain runs through his mouth. He must snap the links with his teeth.

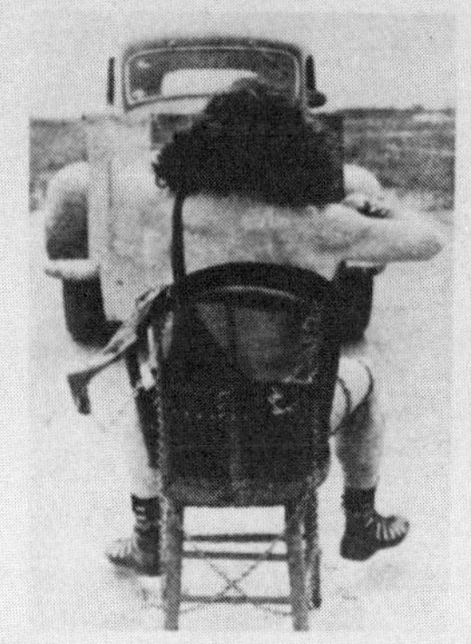

When the bearded escape artist chews through the chain, he snaps the links about his chest by swelling out his deflated lungs.

Thus freeing his torso from the upper part of the chair, he grips the chains wound around his legs; pulls until the links part.

A scant second before the car crashes into the chair, Greenstein shakes off the chains and leaps to safety. He has practiced until he can now

is 56 years old and not quite five feet five inches tall. He weighs 148 pounds and declares himself the strongest little man in the world.

landed on the road with a few bruises. The frame-by-frame action was published in *PIC* magazine, issue of July 25, 1939.

He cooled Leah's ire by loudly denouncing the spate of well-paying suicidal offers that arrived in each morning's mail.

In August of 1938, a German Day rally had drawn a turnout of forty thousand wildly cheering spectators for a parade of two thousand uniformed Nazis. Not in Munich but in Yaphank, Long Island. Joe Greenstein's anti-Nazi battles had begun as soon as Hitler's supporters had attempted to sink American roots. He had an idea of what was coming. "Throughout time, for the Jews it never changes. Fight to live. There is no alternative." The Nazi was a creature of the streets, and there Joe lowered himself to meet them.

He revised and augmented his lectures. In addition to his discourse on clean living, he talked of current events. Pinned to his metal-covered board of 2-inch pine was a caricature of a pig wearing a swastika arm band. The head of the pig was that of Adolf Hitler. After a few choice comments about the German in question, the Atom would take a twenty-penny spike in his hand and smack it through the Nazi pig's heart. The crowd cheered its approval, but certain others didn't think it quite so funny.

About forty years later, Norman Jacobs, Joe's son-in-law, remembered an incident of that time. "I was sitting in their Park Place kitchen with my wife Mary and Leah, when we heard a wild commotion outside. We looked out the window to find Pop mixing it up with four men in the alley. I started out the door to help, but Leah ordered me to stay put. 'Pop will take care of it. But you . . . ,' she warned, 'if you go out, you'll get hurt.' "

When the sounds of battle had subsided, the family went into the alleyway where they found the first combatant sprawled unconscious, his arm and leg broken. The second assailant was discovered in a garbage can, head and legs down, having been folded up and deposited like a discarded sandwich, the can cover neatly on top of him. The third man fled. Joe sat atop the fourth bruiser, putting the finishing touches on him.

The Mighty Atom had spent the afternoon making one of his

anti-Hitler speeches; the Nazi quartet had followed him home and waylaid him in the alley.

"Who are they, Pop?" Leah asked him.

Joe got to his feet, brushed off his pants, and surveyed the men in the alley. "Nobody," he said.

There were dispiriting moments on mornings when he considered the number of well-organized Nazi goons that he might have to go up against that afternoon. At these times there was a passage in the Bible that revitalized him. His was the same secret weapon with which Joshua and Israel had overcome the terrifying horde arrayed against them:

> . . . I will be an enemy unto thine enemies,
> and an adversary unto thine adversaries.
> For mine angel shall go before thee . . .
>
> —Exodus XXIII:22; 23

Shortly after a huge Nazi rally in Madison Square Garden in February, 1939, the Mighty Atom found himself walking through Manhattan's Yorkville German section on a business matter. He stopped in his tracks at a sign in bold letters posted on a building's second floor: "NO DOGS OR JEWS ALLOWED!"

He stared at it for a while before inquiring of a passerby, "What the hell is that?" He was informed that a Nazi Bund meeting was being conducted upstairs. He went across the street to a paint store, where with a three-dollar deposit, he rented an 18-foot ladder. Back he came and opened it beneath the sign. Returning across the street, this time to a sporting-goods store, he purchased a Louisville slugger baseball bat—a "Hank Greenberg Model." He parked it in the doorway beneath the sign.

He went up the ladder, tore the sign down, and tossed it into the gutter. The operation had not gone unnoticed. Several of the Nazis looked out aghast from the second-floor window. The action which followed was in the best tradition of a Popeye cartoon. Before the Atom could climb down from his high perch, the entire Bund assembly had come charging down the stairs into the street.

The Mighty Atom was shaken off his ladder, but he came up bat in hand. They came at him singly and in numbers, frontally and encircling; all to no avail. "It wasn't a fight," Joe said later, "it was a pleasure." He sent eighteen of them to the hospital in various stages of extreme disrepair. He sustained a black eye.

Hauled into court on a charge of aggravated assault, mass mayhem, and so forth, a bedraggled but surprisingly cheerful Joe Greenstein stood meekly and alone before the bench, his only compatriot the "mouse" under his eye. A white-haired judge looked solemnly down as the charge was read. The jurist could hardly believe that the mild, little man before him could have perpetrated such an assault. Then, he surveyed the victims before him, a veritable parade of broken joints, purple contusions, and awkwardly plastered and wired limbs. The battered Aryans filled half the courtroom.

"You mean this little man . . . did that . . . to all of them?" the judge inquired in disbelief.

"Yes, Your Honor," nodded an eyewitness police sergeant. "Them that ain't still in the hospital."

The judge turned his attention to the defendant. "Mr. Greenstein, these are serious charges. Do you have anything to say?"

"Yessir, Judge." The Atom brightened. "*Every time I swung the bat it was a home run!*"

Quietly, the judge inquired of the sergeant what had provoked such a clash. "Them're Nazis, Your Honor," the officer whispered. "They went after him."

"NOT GUILTY! CASE DISMISSED!" The judge banged his gavel.

"But, Your Honor . . ." the Bund's lawyer protested.

"*I said, case dismissed.*" The gavel boomed again with finality, and the judge retired to his chambers.

This battle had had a preview four years earlier. As the nation followed the Lindbergh kidnapping trial of Bruno Hauptmann, and *Lives of a Bengal Lancer* with Gary Cooper made its debut at the New York Paramount, the Mighty Atom was getting ready for his performance at the Arch Street Theatre in Philadelphia. It was January 11, 1935. He had just gone down to the basement of his home at 1480 Park Place, Brooklyn.

"I had to pick up some gear before the show. As I turned . . ." He was hit on top of the head with a bludgeon. "I went down . . . stunned. Then someone put the gaslight out." The place was now only dimly lit by a cellar window. "I heard a voice say, 'A present from Hitler!' " The shadow of a man holding a knife rushed out of the half-darkness.

The blow on top of his head had been a severe one, but it was there that the Atom had his "muscle," that thick layer of interior scar tissue built up after years of pulling trucks and holding back airplanes. At its center, this cap of tissue was a full half inch thick. It had acted as a cushion, preventing him from being knocked unconscious. "I heard quick footsteps and saw the glint of the knife coming down. . . ." He seized the wrist with a deft Jujutsu movement, plunging the redirected blade into the assailant's own chest. He rolled off.

"The blow from the club had weakened me, but when the second man came toward me I managed to land a couple on his face. I broke his nose and maybe his jaw. I could feel the spray of his blood on me."

"Hey"—Joe's sinister whisper carried through the dark—"how are you going to get out of here?" Moments later he heard the sound of glass breaking. The two would-be murderers knocked out a basement window and fled to the street. Joe did not pursue them; his threat had been bluff. He could hardly stand.

When the police arrived, the place looked like a butcher shop. "We found the knife on the cellar floor, then followed the trail of blood out to the middle of the street. Some passers-by had seen the two enter a waiting car with other men." They did not try again.

If in his battles with Nazis and their sympathizers in the United States he would escape serious physical injury, his family in Europe would not be so lucky. Photographs of German atrocities would have special meaning for him; out of 165 of his and Leah's relatives left behind in Poland, all but three would be murdered before the end of the war, swallowed up in an unmarked ditch, the crematoria at Auschwitz, or at a fiery last stand in the Warsaw Ghetto. For Joe Greenstein, the number six million would be a very personal one.

Throughout the latter part of the 1930s the street fights continued with members of the German-American Bund and the fascist

Christian Front. At the back of his lecture podium, Greenstein kept a pail of water with which he doused those vocal Nazi sympathizers he had knocked unconscious. "I didn't want them littering the sidewalk," he said. Once revived, they were sent packing with a swift kick in the pants.

After one of his anti-Nazi speeches caused a near riot at Columbus Circle, Greenstein formed a self-defense group. He dubbed it the Young Maccabees, after the Hebrew band of 165 B.C.E., for whose victory over the Assyrian-Greeks and the cleansing of the Temple of Jerusalem the holiday of Hanukka is celebrated. The group operated in a donated space above Dubrow's Cafeteria off Brooklyn's Eastern Parkway; late into the night the coffee cups would rattle below as, above, bodies were slammed to the mat in Jujutsu practice.

The readers of the Yiddish-American press could take heart from the front-page accounts of the Mighty Atom's exploits against the fascists, almost as one would follow "The Adventures of Terry and the Pirates."

On the pavement or the podium, the Atom would not back off. His private wars had not failed to leave their mark on him. In any number of brawls he had been struck on or near the eyes. On many other occasions, he had burst chains by chest expansion, the flying severed ends flailing up to hit him there. For years he had ignored the advice of doctors who said the cranial pressures attendant to biting feats would damage his sight. When vision in his right eye became blurred, an operation was performed at Brooklyn Eye and Ear Hospital to correct it, but unsuccessfully. The right eye went blind. Specialists advised him to have it removed. He advised them to "take a walk."

When at last America entered the Second World War, he immediately went to Lexington Avenue to the office of the U.S. Army recruiter. Having been rejected for service a quarter century before, he vowed to get accepted this time.

He stood with a parade of apple-cheeked teenagers who chuckled at his presence on the line. Finally, Joe was next and went up to the recruiter, a starched and burly corporal who scrutinized him from behind an olive-drab desk.

"Yes, sir."

"I want to join the Army. Combat duty."

"How old are you, sir?"

"Thirty-two." It was a bad lie; he was two years shy of fifty.

"You've got one blind eye, sir."

"That's okay. I've got another one, a good one. Where do I sign?"

"I'm sorry, sir." The young recruiter shook his head. "There's not much you could do in the Army."

"You don't understand. It's very important. You see, I have family in Europe. I have to help. . . ."

"I'm sorry," the corporal replied, "but thanks for coming in." He offered a handshake.

"Please," Joe begged him, "you need me." The Atom took his hand and applied an infinitesimal amount of pressure; the corporal's eyes widened with the understanding that his bones might be squeezed to jelly. But Joe was not about to.

"I'm really sorry, Pop," he said, "I don't make the rules."

Joe thanked him for his courtesy and left.

Recruiter after recruiter, in all the services, had said the same thing. He was too old and partially blind: he could not go. Tormented by his failure to enlist, he went to see the mayor. LaGuardia, whose acquaintance he had made during his vaude days, and for whom he had campaigned by pulling a twenty-three-ton truck in front of the Brooklyn *Daily Times* building, was a friend whose door was always open to him.

The mayor appointed him to the newly formed New York City Police Patrol Corps. As an auxiliary arm of the Police Department, the Patrol Corps served as a vital aid to a force whose manpower had been badly depleted by the war effort. It also freed other able-bodied policemen for active service. Joe Greenstein was appointed Senior Self-Defense Instructor; rising quickly to the rank of Captain, he taught hand-to-hand defense, imparting his amalgam of Jujutsu and wrestling. He dedicated himself to this position throughout the war years, instructing, giving demonstrations, selling war bonds, and raising money for the allied cause.

- - -

After the war, Joe's forays in search of a livelihood took him farther and farther from home. Traveling in a beat-up touring car loaded with his health products, he cooked his meals out in the open, slept on the ground, and pitched a tent in the rain or cold.

Leah refused to separate herself from him any longer. "I'm going with you, Pop," she said. "It's too much for you to have to do alone." He did not turn her down; she had all the qualities he lacked. She was practical and could handle money, she was a perceptive judge of character. She belonged with him. Their daughter Esther would look after things in her absence.

As Mom and Pop Greenstein worked the fairs and markets late into the fall, they would routinely motor long hours, eating irregularly, catnapping in the front seat, to set up shop on some windblown fairground in the chill of morning. Often they would work the day in rain-soaked clothes.

Out of curiosity, their son-in-law Norman decided to help out and go pitching with them. It proved too much for him; dog-weary, he quit after four days and took the train home.

Yet Mom and Pop Greenstein accepted this life; even when the car broke down, or money ran out, or the road presented real danger. On the way to upstate New York, their right front wheel sheared off in a pothole, the machine yawed and turned over. Leah was doused and burned by battery acid; Joe pulled her free through a window. Sometime later, while they were driving on a downhill grade of Jim Thorpe Hill, Pennsylvania, she became alarmed when she noticed that he was driving too fast, his jaw set, his hands clenching the wheel.

"Pop . . . ?"

"Hold on," he said quietly, "the brakes are gone." He tried the handbrake, the friction set the wheels afire. Finally he stopped the overloaded car by ramming a house.

Through it all, Leah could always be relied upon to cheer him up with a wisecrack. Each time she returned home with presents for the children.

"There is no reason why any woman cannot physically satisfy any man," Joe Greenstein said. "But physical attraction soon cools.

Spiritual attraction. That never fades. It gets deeper." This life on the road made them closer still.

Ultimately, they needed a multipurpose weatherproof vehicle, a combination motor home and traveling lecture stand. As usual, there was only one way for Joe Greenstein to get what he needed—he would have to make it himself. He bought a new Dodge truck chassis and wheeled it around to the backyard of his home on Park Place. There, he set about building with his usual flurry of enthusiasm, laboring like a Noah over his land-going arc; he constructed a shell of solid oak and meticulously covered it with heavy sheets of hammered tin. At last, after three months, it was finished.

The Atomobile was a beautifully crafted job complete with sleeping accommodations for two, cooking facilities, electrical outlets, air vents, porthole windows, and a rear section which opened into a lecture platform containing a sound system, lights, and doors displaying his photographic museum.

The Mighty Atom had a way with animals that approached sorcery, the alchemist's apprentice being a wiry black-and-white terrier named Patches that he trained not to fear fire. This canine exhibitionist would pull on his pants leg for him to light matches, so that she could extinguish them with her paw. Joe could always gather an audience by having Patches gleefully hurl herself through a fiery hoop.

Joe kept the pick of her first litter, a white male with a brown diamond on his back, named Lucky. The Atom held these animals in such affection that he customized the truck for them, building special alcoves alongside the driver and passenger seats to accommodate them.

A young woman with fond memories of the menagerie in the front seat is Joe's granddaughter Elaine. When her father lost his life in a tragic road accident, Elaine and her brother Robert came to live with their grandparents. "There never was much money in the house," she recalled. "But no child ever lived in a home with more warmth and emotional security." Often the whole crew would go pitching in the Atomobile, Joe behind the wheel with Patches on his left, Leah in the passenger seat, and Elaine and Lucky beside her in the right alcove; a chorus of singing and barking as they went down the road.

The extent of the dogs' attachment to their home was proven years later when Lucky reenacted the ultimate cliché of all faithful-dog stories. While they were pitching in Pennsylvania, their mutt was lost in a snowstorm, and after searching for him for a day Mom and Pop returned home emotionally spent. Several weeks later Lucky appeared on their doorstep, having trekked home two hundred miles through the snow.

Patches and Lucky would be the Mighty Atom's constant companions, tagging at his heels, each sleeping at the foot of his bed for their life span of eighteen years.

Mr. and Mrs. Atom and their dogs often took the road south. Leah would sit under a big umbrella against the sun while Joe lectured and demonstrated. Many southern towns had ordinances to bar pitchmen, but in many of these places he was welcomed, often by the mayor or chief of police. From the beginning, he held most other pitchmen in contempt. "If you can't do any good," he would brace them, "don't do any harm."

Unlike the others who may have left town at the point of a sheriff's boot, the Atom was himself an honorary lawman in many of these localities, and brandished a collection of admiring letters and commendations from officials:

> To Whom It May Concern,
>
> This is to acknowledge the valuable work rendered the Physical Training Department, this headquarters, by Professor Joseph Greenstein, better known as "The Mighty Atom."
>
> While on this field, Professor Greenstein performed on several occasions for education and entertainment of the enlisted and Officer personnel. His performances were received with great enthusiasm and the educational value of his appearances as it related to the physical training program of the army is really immeasurable. . . .
>
> This office and all members of the Physical Training Staff heartily

endorses Professor Greenstein for an important part in the Army Physical Training program.

LT. C. W. H.
Physical Training Branch
Army Air Force Technical School
Seymour-Johnson Field, North Carolina

Dear Sir,

Our students and teachers thoroughly enjoyed your performance and were amazed at the feats of strength you performed . . . the wonderful remarks you made to the students about how to grow strong, take care of their bodies, and certain moral attitudes. We hope to see you again. . . .

DUNN PUBLIC SCHOOLS
Dunn, North Carolina

In traveling, the Atom's first order of business was often a charity affair for civic benefit, or self-defense instruction to the local police. On one such occasion, he gave a three-hour show in Moonachie, New Jersey, capping the event by towing Moonachie Engine Company Number One, an Ahrens Fox pumper with a complement of firefighters on it, down the street Samson style.

One southern town that was off-limits to pitchmen was Columbia, South Carolina. After talking to several local dignitaries, Joe was able to arrange a meeting with the mayor and chief of police. Though impressed by his demonstrable qualifications as a lecturer and strongman, neither seemed inclined to let him work Columbia.

Leafing through the Atom's book of congratulatory letters, the chief spied several commendations for self-defense instruction to police departments. He examined a particular newspaper account with a critical eye. "It says here you instruct in disarming a man with a loaded revolver." Despite the documentation which attested that the little man had indeed disarmed men with revolvers on numerous occasions, the chief was dubious. He suggested that the Atom make such an attempt against his local firearms instructor. To simulate actual robbery conditions, the approach would be from behind with weapon at close range. To make things interesting, the Atom volunteered to wear a blindfold, accepting the challenge with

FRIENDSHIP
ENGINE & HOSE
ENGINE CO. NO. 1
MOONACHIE, N.J.
AHRENS-FOX

the provision that should he succeed, he could do business in town.

He was led into a barn and stood at its center. He heard footsteps coming up behind him, and reacted at the first twinge of the pistol jammed into his back.

Outside, the waiting lawmen heard a gunshot. The strongman emerged with the smoking revolver and blindfold in hand. He had begun to raise his hands but whirled, deflecting the weapon with a quick elbow. The live .38 bounced across the barn floor, discharging into a wall With a deft reverse joint lock the "assailant" was subdued, having made the mistake of signaling his exact position with the gun barrel.

From that evening on, Joe Greenstein pitched in Columbia with the blessings of the law. After several days of shows and lectures he added one more official commendation to his collection;

> To Whom It May Concern,
>
> This is to certify that Joseph, the Mighty Atom, Greenstein put on a demonstration of his scientific skill and strength, both at the police department and the city hall.
>
> I witnessed these feats and verify the fact that they were done exactly as he claimed, though seemingly impossible.
>
> W. H. R.
> Chief of Police
> Columbia, South Carolina

Southern byways offered their share of curious encounters. On his way through the Carolinas, the Atom stopped at a rural barber shop to freshen up before his evening lecture.

"You're that Mighty Atom fella, ain'tcha?" the barber pursued conversation. "I hear you're gonna bite a nail in half tonight."

The strongman nodded matter of factly.

"I don't believe it. Nobody could do that."

"Bring a spike to the lecture," Greenstein suggested, and reclined in the chair.

That evening, halfway through his program, the Atom noticed the barber, now somewhat in his cups, weaving toward the front row.

"Here ya go," the barber interrupted. "I brought ya the spike. But I still don't believe it. And I don't think anybody else does either."

After a few barbs aimed at the drunk, the Atom exhibited the iron to the crowd. Then he bit down. Nothing. The spike was hard as diamond.

"See, I told ya," the barber exclaimed, "*impossible!*" The sizable crowd began to buzz with skepticism, and Joe became concerned. His reputation was his livelihood; word would spread fast if people believed him a fraud. A traveling strongman with a dubious reputation could never put dinner on the table. He did not hold back. He bit down again . . . crack . . . crack . . . . When the spike was finally in two, blood was running from his mouth. He had lost three teeth at the price of reputation. He returned the severed iron to its amazed owner who examined it carefully.

"Son of a gun. I figured you switched nails, or somethin'. But this one's mine. No doubt about it. See there . . ." He indicated identifying scratch marks on the nailhead. "That's fantastic."

The audience applauded enthusiastically, and the tipsy barber departed back into the crowd mumbling, ". . . *and I even tempered it, too!*"

In his health lectures the Atom often cited the Kosher laws, having long ago discovered the fundamental truths which they embodied. The Mosaic Dietary Code, which was the first to suggest that you are what you eat, that the diet which pollutes the body will also pollute the soul, was to the Atom not dogma but simply clean living.

"In the Book of Genesis, animal blood is forbidden to all the seed of Noah," he explained. "Now, we know that blood carries off the impurities in the muscle tissues of animals, that it circulates germs and spores of infectious disease. . . . Moses himself forbade the eating of animal fat, thousands of years before anyone heard the word 'cholesterol.' Now we discover that cholesterol, increased in the body by fat consumption, may result in hardening of the arteries and coronary heart disease. . . . Pork was also forbidden millennia before the discovery of the microscope and the identification of trichinella spiralis, the invisible parasite that causes trichinosis . . . . Shellfish, which we now understand may contribute to typhoid and hepatitis, were also prohibited, and they are the highest in heart-killing cholesterol as well. . . . The Mosaic law even extended to isolation of those with communicable disease. You may not know it, but as Commander-in-Chief of the Army of the American Revolution, General George Washington urged his troops to emulate the Israelite's Biblical laws of cleanliness. His writings may be found in the Houghton Library of Harvard University . . . ." Joe often explained the wisdom of the Kosher laws in detail to his interested and predominantly gentile audiences, but he sought only to influence his listeners toward a healthful way of life, and to make his living.

"I am vehemently opposed to those who seek to convert others for their own religious purposes. I have vigorously resisted such persons for my entire life."

There were many who were not as considerate of him. As a learned man who studied the Bible and Talmud, he would have been quite a coup for the missionaries who never stopped trying.

He and Leah had been working at a Pennsylvania fair, and business had been rather poor that week. On the grounds was a large circular tent where a self-appointed preacher of the more strident variety conducted his sermons of hell-fire and damnation. Sometimes, after he had preached himself hoarse, he would walk over to the Atom's lecture stand for another pot-shot at his favorite target. Always, after engaging Joe in conversation, he would quickly move the subject to religion.

"Now, listen, I just told you, I defend your right to your own beliefs." Joe was getting a bit miffed. "You must either have the decency to respect mine, or leave me alone."

"If you will only accept what I say, Greenstein," the man went on, "wonderful things will happen to you. Mark my words."

"Yeah, sure. Have a nice day."

The man raised his eyes to the heavens. "Show him a sign, Lord!" his voice thundered. "*Show him a sign!*"

An hour later, Joe began his next lecture, and the week's mediocre business abruptly improved. The crowds doubled and then tripled, people began buying his products by the armload. By the end of the afternoon, he was sold out of stock; never in all his years of working had he seen such a windfall.

When he finally stepped down from the platform, he pondered his unusually good business, then remembered the preacher's words. The thought irked him.

"Joe," Leah said, "something's not kosher here."

"I'll be back in a minute, Mom," he told her, and made his way to the preacher's tent. Every seat was filled. As he walked in, the speaker's eyes widened and he called out, "Welcome, brother."

Joe looked around and noticed immediately that all those present had Mighty Atom products by the bagful. Now it was clear that the good Reverend had instructed his devotees to buy Joe's products to gain a convert.

Joe walked down the aisle and stopped in front of the man's pulpit; smiling, he reached into his pocket, took out a wad of bills, and tossed it in the air, the bills fluttering into the aisle. "No sale!" Joe said, and walked out.

In the late 1940s, the Atom took savings of $18,000 and founded a health resort where he could put his ideas into action. Where better to advocate health and vitality than in bucolic surroundings, far from the noise and dirt of the city? In upstate Woodridge, New York, he purchased a seventy-acre tract of land with a twenty-three room hotel, four apartments, and two bungalows and established his Panoramic Health Farm. The plan was that he would run the place all summer until Labor Day, when he and Leah would take their traveling lecturemobile south for the winter. Leah had her doubts about an entire hotel being a simple husband-and-wife operation. With his usual enthusiasm, Joe approached the idea as if it were nothing more difficult than a mom-and-pop candy store. He was not to be dissuaded. He put down the cash and took title.

Of panoramic view there was plenty, of water there was none. A week after he took over, the sole spring went dry. The seller (who had paid $7,000 for the place a few years before) had neglected to mention the periodic problem.

After several weeks of carting water from town in barrels and cans, Joe went to the library to research the problem. He found reference to individuals who did nothing more for a living than discover water for such unfortunates as himself. These "dowsers" were supposedly gifted with the ability to sniff out H_2O with nothing more than a divining rod.

In Pennsylvania, he found just such a pair of "water smellers." Dressed in black, and possessed of an appropriately mysterious manner, the pair immediately made him suspicious. Instead of your everyday divining rod, they did their dowsing with an upended pliers. At last, they stopped at a miserable patch of weeds and pronounced with finality that they had found water. As the pliers were jiggling with wild and spastic enthusiasm, he took them at their word. They returned to Pennsylvania, his cash in their pocket.

He called the local well driller, a Sicilian who arrived on the scene with well-founded cynicism, as in the very place where the dowsers had predicted water, he had already dug a dry well for the previous owner.

Now out a couple of hundred dollars for a pair of sham water smellers, and a good chunk of his life's savings for the health farm itself, Joe nevertheless did not despair. He would find water . . . or throw himself off the nearest bridge. Somehow, a wet death seemed almost pleasant under the circumstances.

He took a large flat rock and a sixteen-pound sledgehammer, placed the rock on the ground in various locations of the property, and smacked it soundly with the hammer. He reasoned that if there were water below somewhere, there might be an underground echo or other indication.

He found an area that responded. The more he hit the rock, the more he became convinced that this was the spot. Immediately, he summoned the well driller.

"Here?" The man was not encouraging. "Are you kiddin'? I already drilled right here, too. I didn't find enough water to rinse out my mouth."

Joe could not be dissuaded, and after signing a contract guaranteeing payment, he told the man to go to work. The bits were sunk into the ground, and there was nothing. The driller looked at Joe blankly. "Dig deeper," Joe ordered, and gave him more money. Nothing. "Deeper!" He doled out the cash from the piggy bank.

At last came a gathering gushing sound, and a geyser of water sprayed high into the air coming up at the rate of seventy gallons a minute. The little Sicilian crossed himself.

"How did you know?"

Joe shrugged.

"Mister Atom"—the man pointed heavenward—"you got somebody upstairs."

With a bit of borrowing and some juggling of finances, Joe fitted a pump on the site, made a small lake, stocked it with fish, and put two boats on it. He enlarged the approach road, constructed another two-story guest house, and built a pool. He was working seventeen hours a day and by now his investment had gotten out of hand, about $55,000 out of hand. He began alternating one week at the Panoramic, one week of pitching night and day to try to pay the previous week's bills.

The Panoramic Health Farm was no mom-and-pop candy store; the clientele was an eccentric and demanding bunch. For the first time in her life, Leah started visiting doctors. The diagnoses were the same: overwork.

After a decade, rather than have the Panoramic Health Farm kill them, Joe sold out for such a disastrously low figure that he needn't have drilled for water; he ended taking a bath in his own money. He gladly returned full-time to the life of a pitchman.

That the strongman had not been completely forgotten was evidenced by a little brown filly who ran the American flat tracks as the half century dawned. "The Mighty Atom," by Fighting Fox, out of Chamonix, flew the colors of the Wheatley Stables. Winning her share of races between 1951 and 1954, she was retired a year later and bred two foals. Though her racing record was laudable, her prolificacy never matched that of her namesake.

10.

SLIM THE HAMMER MAN

In Gilbertsville, Pennsylvania, on a muggy Saturday night in the summer of 1955, Joe Greenstein berated the youthful crowd at Zern's. "You drink, you smoke, you destroy yourselves with ignorance. . . ." He held up a steel spike. "How is it that only I, an old man from among the lot of you, can bend the sixty-penny spike that I hold in my hand?"

A voice came out of the crowd: "You ain't talkin' about me." The voice belonged to a man in his twenties who towered above the others. Joe had seen him for years, first as a gawky youth standing silently before his platform; the boy had grown to a giant of a man. They had in all that time never spoken, and Joe did not know his name.

"What did you say?" Joe asked.

"I said you ain't talkin' about me; I ain't like these clowns."

"Come up and show me," the Atom invited.

Slim Farman walked toward the platform; he was 6 feet 6 inches, and 210 pounds, yet wiry and agile. He did not smile and looked like a solemn L'il Abner. Ignoring a congenital heart murmur, he worked as a stonecutter and foreman at Gil's Quarry, Fairview Village, Pennsylvania. Wielding a sixteen-pound hammer nine hours a day, sometimes, breaking and hauling one hundred tons of stone each day, he developed a body as hard as the stone he pulverized, and massive, cable-strong arms. Farman took himself seriously. He smashed his way out of poverty to middle-class status with nothing more than a sledgehammer.

Farman strode up onto the platform and stood in the glare of the lights. "I've seen this man for years, seen him do things," he thought to himself, "but he's half my size, twice my age, and he's a Jew." Farman, not having known any Jews in his rural upbringing, had acquired the stereotypical notion that they did not excel in the physical. Slim Farman was the quintessential gentile; a Jew was a tailor or a man who smelled of books and no tiny old Jew was a better man than he.

"Can you bend this?" the old man asked again and offered him the spike.

"If you can, I certainly can." Farman accepted it.

"Go ahead." The old man watched with curiosity. Farman took it in his hands and went at it. Without technique or experience, but with visible determination he succeeded.

The Atom was impressed. "Have you ever done that before?"

"No."

"Then how did you know you could?"

"Because I said so."

"Only words."

"No," said Farman, "I'd rather die than break my word to myself."

The Atom inquired if he could do any other feats of strength. Farman told him about one that he had only begun as a wager, the leverage lifting of a sledgehammer. Anyone who has ever tried to hold a broomstick horizontal by gripping its very end will know how exceptional the same feat would be with a sledgehammer. Joe asked to see him do it. Farman went home and got his hammer. The Atom watched carefully as Farman lifted the twelve-pound hammer which was lying on the floor up into a vertical position while holding only the end of its handle. After the lift, the Atom tried himself and failed. "In all my years as a professional, I have never seen another human being do that," he said. "You have something here, my boy, and should pursue it seriously."

Over the next weeks, Farman came back to Zern's with a changed attitude. He was not impressed with himself for having bent a spike, but with Greenstein who at sixty-two years of age tore horseshoes apart with less effort.

"The old Jew is a better man than me," Farman admitted to himself. "This man knows secrets."

"Teach me, old man," he said.

"I can't, Slim." The Atom smiled mysteriously, and said nothing more.

Though put off by the rebuff, Slim Farman nevertheless returned every weekend.

"*Eat dead food . . . and you're a dead duck,*" the old man warned. "If you succeed in obtaining the necessary vitamins and minerals from the food you eat, then you're alive. If you don't eat well, eliminating the nutrients by ignorance, then you're eating dead food.

"I do not care what food looks, smells, or tastes like. I care only about what it can do for me . . . or to me. I advocate a low-fat, low-sugar, low-salt, mainly vegetarian diet, with some fish or poultry, fruit juices, and only an occasional egg or piece of red meat."

For the first time, Slim Farman began to think seriously about his health. He followed the Atom's dietary regimen. He gave up liquor. After a smoking habit of more than a dozen cigars a day, he quit cold, keeping a couple in his pocket for discipline. "You ain't got me. I got you."

Weeks turned into years, and he always came to see the old man. "Slim, after your hard day's work, you always go to see that old guy up at the market. What for?" his co-workers at the quarry asked in puzzlement.

"You'll think I'm crazy," he answered.

"Tell us anyway."

"I feel alive, physically and mentally recharged just being in his presence," Slim said.

On a warm Saturday night, Slim watched again as the old man twisted a horseshoe as if it were a pipe cleaner and dropped it matter-of-factly at his feet. After his lecture, the Atom stepped down from the platform and the big man stood holding the iron in his hands.

"Old man, I'm stronger than you." He shook his head. "*I know I am. And I can't do it!*"

The Atom offered only his arcane smile.

"If I figure out how you do it, will you tell me if I'm right?"

The old man nodded.

Farman thought of little else: it ate at him, confounded him, until one night months later he awakened from sleep and sat bolt upright.

Returning again to Zern's, he sauntered up to the Atom.

"You don't use your body. You use your *mind*. That's it, isn't it?"

"Yes."

"*Now* teach me, Joe."

"I can't, Slim."

"Why not? What's wrong with me?"

"I can't teach you what you already know."

Slim looked at him with confusion.

"Slim, remember when I asked you how you bent that spike? You said you'd rather die than fail."

Farman was overwhelmed at his recalling their conversation word for word, now years later.

"You'd rather die than fail," Joe said. "You know it already. The rest is practice."

"But I don't *feel* like I have it, Joe."

"You don't feel the hair on your head, either, but it's there. This is not something you feel, Slim, it's something you *live*."

Slim began to train in earnest as the Atom's protégé. When he first began the hammer lift, Slim believed that his best could be sixteen pounds with either hand. The Atom demanded that he lift eighteen.

"I can't."

"How do you know you can't? Do your arms fall off?"

"No."

"Of course not. Your instinct for preservation makes you quit. But you can go beyond, you can train yourself not to quit. How bad do you want it?"

- - -

"Slim, my boy," the Atom said time and again, "never inhibit or limit yourself by the seemingly impossible." An incident in the quarry raised Slim's level of cognition.

While he stood atop a loading bin, freeing crushed rock with a probe, a moving hole suddenly opened beneath his feet and the lake of stone swallowed him. Conscious, he lay under thirty tons of crushed rock, hands cupped in front of his face, unable to move any part of his body. There was no pain or great pressure, like being covered in sand at the beach; he had become part of the stone. He thought it out clearly. He couldn't move; there was no way out; no one had seen him go; he made his peace. After several minutes, the air supply ran out, his consciousness ebbed away. He knew he was dead.

Minutes later, the load of rock suddenly shifted and moved again. From the overhead chute in a shower of stone he was deposited onto a conveyor belt in front of his startled men. His eyes opened, he rose on shaky legs, bloody, his stomach punctured by a broken rib. They stared at him in disbelief.

"What're you lookin' at?" he said, "get back to work." Ignoring their suggestion that he get to a hospital, he limped back up onto the bin to free the stone which had jammed again.

Like the Atom's walking away from a bullet, Slim's survival was real, and it expanded his new way of thinking. He went back into training. Under the Atom's tutelage, he progressed to twenty-six pounds with either hand, a combined leverage lift of fifty-two pounds at 31 inches, with 1,612-inch pounds of pressure on his wrists.

"How bad do you want it?"

Slim's hands acquired small bumps between thumb, forefinger, and wrist, souvenirs of the numerous occasions on which bones stress-fractured from the enormous lift pressures. Broken bones never deterred him from a lift. He completed, allowed the bones to heal, then increased the poundage. His chromed hammers were affixed with a center bolt to accept plates of added weight. At night, they rested in a red-velvet-lined case. These hammers became an extension of his very being. There is a proviso in his Last Will and Testament that the polished oak and chromed steel hammer which he retired after eighteen years' service in the quarry is to be buried with him.

Slim came to accept astonishing exhibitions as part of his education. He had been close to the Atom for some years when one evening, between lectures at Zern's, he and Joe sat beside the truck continuing their conversation about athletic potential.

"Slim"—the Atom pinned him with his cryptic gaze—"there is no such thing as an involuntary response."

"The heartbeat?"

"The mind can control the working of any organ, any part of the body. The brain rules the heart; not the other way around. I can regulate my own heartbeat, my pulse—I can commit suicide by breath control right now if I want to."

"I believe you, Joe."

"Of course you do. But mere belief is not enough. You must know. You must see for yourself."

Slim knew that the old man was up to something. "What are you going to do?"

"Don't worry," Joe said, "just watch."

There were about fifty people gathered before the Atom's platform when he went up again.

"Is there a doctor or nurse present?" he asked.

"I'm a registered nurse." A young woman stepped forward and offered her assistance.

"Ladies and gentlemen," the Atom said, "in a conversation with my friend Slim, I told him that the heart is matter, and that mind rules matter. To demonstrate this I am going to stop my pulse." The audience waited for the punchline, but there was none; he was serious.

He sat down and the nurse placed her two fingers on his wrist.

"You've got a pulse like a horse," she laughed.

The old man did not speak, but looked off into the distance; her smile suddenly disappeared.

"It's starting to slow down," she said, startled but continuing to measure with exactness. "Even slower now."

His face became colorless.

"My God, I can't feel it. It's stopped!"

After about ten seconds, his pulse slowly returned, and five minutes later he was back to selling liniment. After the lecture, the old man stood with his student beside the truck.

"Now you know," he said.

One of the most interesting reactions to Joe Greenstein's demonstrations took place in September of 1935 when he and his daughter Judy made a six-week auto tour of the rural South. They camped for a while near Smithfield, North Carolina, where the naturopath and his products caused a stir. It quickly became evident that this man was "a healer," a title highly regarded in the backwoods. Prophetic looking as he was, replete with shoulder-length hair, performing superhuman deeds, possessed of the ability to stop his pulse and heartbeat at will, he caused some of them to take him as Jesus Christ himself. When whole families started following him around for no other reason than to touch his clothes, or kiss his hands or feet, he knew it was time to pack up and leave.

The Atom always claimed that his mind control could be transferred, and one evening he proved it. At Zern's, with Slim near the platform, the Atom called for a volunteer, and a man of medium build in his forties stepped up.

"What's your name, sir?"

"Lloyd."

"Lloyd, will you bend this sixty-penny spike for me?"

The man laughed bashfully, tried, and got nowhere.

"Lloyd, will you trust me?"

"Yes."

"Will you do as I say, knowing that I would never hurt you?"

"Yes."

"Good. From now on your name isn't Lloyd. It's Sam."

"Okay."

"What's your name?"

"Sam."

The crowd looked on as for several minutes the Atom implanted his suggestion.

Courtesy *Norristown Times Herald*

"Sam, yesterday we were walking in the woods. You remember that, don't you?"

"Yes."

"And we found a long thin piece of wire."

"Yes."

"And I bent the wire back and forth so easily."

"Yes."

"And because it was so simple, I gave you the wire. And you bent it so easily."

"Yes."

"That it took no effort. It was a weak, thin thread of wire."

"Yes, I remember."

"What's your name?"

"Sam."

"Sam, I cut off a piece of that wire and I have brought it here with me today. *Sam, you are capable of bending this wire.* People call it a spike, but it's nothing. A piece of wire. I've wrapped the point and head in a bit of rag. It will not hurt your hand. You will feel no pain. Your muscles will obey your mind. You have bent this wire before, you can bend it now . . . *and you will. You are stronger than any piece of wire on earth.*"

All the while he affixed "Sam" with his unwavering eye. "Aren't you?"

"Yes, I am."

"Now Sam, take the wire and bend it."

The old man handed over the 6-inch-long, ½-inch-thick spike, and "Sam" bent it into a 90-degree angle as Slim Farman and the rest of the audience watched with mouths agape.

"Thank you, Sam," the Atom said matter-of-factly.

One minute later, the old man had brought him out of the suggestion by the same means, and without ill effect, dispelling the last of it with a light tap on the chin.

"What's your name?"

"Lloyd."

"Lloyd, will you bend this for me, please?" he said handing over another identical piece of iron. The middle-aged man tried his best, and after failing to budge it, stepped down.

Slim saw the Atom do this on several occasions, and these same subjects attested that they have never been able to succeed before or since.

Though the Atom demonstrated such potential within each man, he was very cautious as to whom he seriously instructed, and always concerned himself with the use to which it might be put. At one time, he had taken on a young boxer named Georgie Small. Small was an apt pupil, gifted with natural talents. As Joe traveled pitching, Small accompanied him, training, and absorbing his lessons. Small made his way up through the pugilistic ranks, until his promising career ended when his opponent, a young boxer named Laverne Roach, died after being knocked out.

The story of Georgie Small appeared in many newspapers and magazines of the period, among them a *True Story* article of February, 1962, "Death at Arms Length." The Atom never trained another boxer. Forever after, he wondered if perhaps he had trained too well.

■ ■ ■

With his disciple there could be only complete trust. Joe Greenstein had become Slim Farman's father, teacher, and guide, the only man alive who knew the way Slim had to travel. "In this world, there's no one like him; he's the King," Slim told anyone who would listen. Slim had a loving wife Shirley and five beautiful kids, but after his immediate family, "Joe Greenstein is the only friend I've ever had," he said. Shirley Farman confided, "I've never seen my husband show physical warmth toward another man. But when he sees Joe, Slim unashamedly gives him a kiss." He visited the old man every Father's Day.

At Joe's encouragement, Slim began performing at weight-lifting and wrist-wrestling championships. His lifts escalated to world records. By the summer of 1975, "Slim the Hammer Man" was ready.

"Prepare your equipment," the Atom said.

"Where're we going, Joe?"

"Madison Square Garden."

The Atom had been invited to perform at Madison Square Garden. After his own exhibition, he strode from the center ring to a roaring, standing ovation, and Slim the Hammer Man stood before the crowd.

A half hour before, in his dressing room backstage, he had warmed up and then attempted the leverage lift that he was to perform to break his own world record; a pair of twenty-eight-pound hammers, 56 pounds on 31-inch handles, with 1,736-inch pounds of pressure on his wrists. They lay horizontal on the floor. He set himself, tried to raise them and failed. He made a second attempt, but could not hold the weight. He tried six times to no avail; on the last attempt he could hardly move the weight an inch off the floor, and in doing so he strained both his wrists. Slim grimaced from the pain, and his son, Larry Jr., looked on with concern.

"Dad, it's time. I have to take your hammers up to the ring now. How much weight shall I take off?"

"Leave the hammers like they are."

"But—"

"I told them I would break the record."

"Dad, you're hurt."

World record leverage lift: 56 pounds on 31 inches—1,736 inch pounds. All photos by John Strickler

"Never mind, son. You just take those hammers up there."

Slim Farman faced the crowd of twenty thousand alone. He peered at the mirror-bright hammers, the oak and steel objects which he had loved and cursed for so long, that had broken his wrists twenty times. He passed his hand over the steel of the hammers shielding them from his view with what appeared a theatrical flourish. It was his unique methodology for mind over matter:

"When I put my hand over the top of a hammer, I can no longer see the steel. I move my hand over it and for me, the top of the hammer disappears. Then I pick up the stick."

Twenty years of training were condensed into a millisecond. The energy exploded inside him; he let it go. The horizontal hammers came up off the mat as if by levitation. Pivoting them up, he held them out vertical at straight arms' length. Slowly, without moving an arm or bending an elbow, he lowered the weighted chrome monsters until they touched his head, then he lifted them vertical again. With a pair of 16-pounders he repeated the feat, arms extended out to the side, then again with a sharp weighted twenty-pound woodman's axe, which threatened to split his head if he could not raise it back up. For twenty minutes he performed, the audience reacting with hushed fascination—then, ovation.

The Mighty Atom stood framed in the rear exit archway, nodding his silent approval. His spiritual offspring had arrived.

All around him the world was spinning, changing; only Joe Greenstein seemed to remain the same, as if fixed in place. He grew older, but did not age. "Retirement from activity promotes death," he said. The inner resources which dominated his personality were undisturbed.

In street clothes, he appeared a little old man. But in his Samson's costume, with his white hair to his shoulders, he took on an otherworldly countenance. The gaze of his one blue eye beamed the power of his presence. The life-force was ever-elastic, ever-flowing, unabated.

The everyday aspect of the Mighty Atom's feats could not be underestimated. The Houdini maxim that "magic is practice" parallels the Zen "A used doorstep never rots." Talent left to wither will disappear, while that which is practiced will flourish. The octogenarian Atom ceased breaking horseshoes for economic rather than physical reasons. "I used to buy them in big barrels for three dollars," he said. "Now they're three bucks apiece. Who's got that kind of money?" The most contemporary touch about him was the 1961 white Chevy station wagon that he drove to and from Pennsylvania Dutch country on weekends.

In a changing urban environment, the old man was preyed upon by the muggers who stalk the helpless aged. There were mugging attempts made upon Joe Greenstein, each ending in near tragedy—for the muggers.

After visiting a friend, he found himself in a darkened building hallway, and took no notice of the two shadows that fell in behind him. Suddenly there was a strong arm around his neck, a cheap pistol held against his face.

"Give up what you got, old man."

The Atom was startled, but not cowed. The second thug brandished a knife. "Quick, you old mother, or I'll cut your head off."

The Atom reached up and deftly snapped the wrist behind the gun hand, breaking it with a pop. He dropped to one knee and executed a shoulder throw. The stunned youth was hurled upside down into a doorway, exploding the silence with a crash, and lay motionless.

The second youth backed into a vestibule, frozen with fright. Greenstein dismissed him with disgust. "Go, sonny. Run away." Sneakers flapping, the youth ran for his life.

Time and again, he disarmed them of knives and pistols. He broke their bones. The Atom's personal anti-mugger campaign was, on occasion, featured on the local television news.

"I wear a beard for two reasons," he said, "for religious purposes, and to tempt muggers. Most old people have to endure abuse, violence, and fear. I don't. Let them try to rob me. Let them try to pull my beard. I'll pull their arms out."

The Atom continued on year after year, he and Leah, lecturing and selling his products from the old truck, performing his feats of strength as part of daily routine. Public interest in him increased with the passage of time, sparked by his seemingly inextinguishable

powers. The messages he had delivered for a lifetime were now coming into vogue. By standing still, he had become modern.

Shortly after New Year's, 1974, Leah entered the hospital after suffering a mild stroke. Always surrounded by children, she had the attitudes of a woman half her age; no one thought of her as the seventy-eight years she actually was. She tried to calm the family's fears: "It's nothing, nothing, don't worry." She was released briefly from the hospital but readmitted after another stroke. On January 28, 1974, she died.

Joe lost his mind; he was semi-conscious on and off for weeks. He raved so at the funeral that there were those who never forgave him for this display. But he was as much in mourning for himself as for her. He had lived in a world of dreams all his life; when the frightened little boy needed a strong man to help him, he became that man himself. On every occasion he and Leah had survived, survival was his last illusion, and now that too disintegrated. He talked of suicide.

"Joe, she was seventy-eight years old. . . ." Friends tried to help him get a grip on himself.

"I never thought she could die." He peered at them wildly. "It never entered my mind . . . never."

He was lost without her. Now there was no one to love him, no one to make him believe, no one to show off for. The source of his strength, his right hand, was gone.

11.

THE DAYS ALONE

I was going to see the old man. It was May, 1975, a bright, warm, sweet-smelling day. At the garage in Brooklyn, I picked up my 1951 Mercury Coupe, a modern classic, shiny black with jumbo white balloon tires. It had a small propeller on the hood ornament which spun and emitted a spitfire whine as it went down the road. The Merc was a black torpedo, a machine with presence, the kind they throw bodies out of in gangster movies.

I drove it down Coney Island Avenue, past Kosher chicken stores, "The Friends of the Jews" (a storefront for religious conversions in a constant state of vandalization), the Louis Specter Frame Shop, a couple of funeral parlors, and all the other surviving landmarks of Brooklyn in the twilight of its Jewishness. I had watched too many of her shining princes, Bar Mitzvahed valedictorians, intermarry into Delaware.

I thought it fitting to drive my 1951 Mercury to see the Mighty Atom. I like old things—they have substance, they've been tested. I cruised over struts of shadow on the floor of the Brooklyn Bridge. I moved the big Merc through the Holland Tunnel and my face caught the pillow of daylight and the May breeze as I entered New Jersey. The propeller whining, I headed down the ribbon of Route 22 South, past Fairyland Kiddie Park, Marine Corps Recruiting Posters ("Quality, Not Quantity!"), myriad gas stations (51.9¢ including

Spielman

tax). The low rumble of the flathead V–8 made a joyous buzz in my ear as it carried me through the rolling green of Pennsylvania Dutch country.

I swung the car off Route 73 into Zern's Farmer's Market. The cavernous building and outside stalls of Zern's covered several acres. Here, amidst the canned goods and children's wear, antiques, tools, model railroad trains, beer-can collections, fruit stands and purveyors of smoked cheese, the Mighty Atom was an attraction, an institution accorded the honor of working rent free.

Joe was waiting beside his truck. The battered vehicle was moored in its stall. The Atomobile was now on its third engine, a rolling museum piece, a remnant of his American odyssey, a motorized imitation of Volanko's van.

Inside amongst the junk were bins of spare parts for this and that, pots and pans, a couple of battered radios from the 1930s. In the innermost recesses of the vehicle one could find the artifacts of his career. No. 5 horseshoes which he had bent into unlikely shapes, a plethora of hand tools in use since the turn of the century, an American Indian drum picked up in Texas in 1916, its painted feather now faded and barely visible.

In the afternoon, farmers came by in family groups to say hello. Some talked kindly of his wife's memory. He thanked them, turning his head and weeping unashamedly. This was no purely business relationship, these were his friends.

A mechanic came over to buy some of Atom's soap. When told that it might be the old man's last season, he promptly took out a twenty-dollar bill and bought a whole box.

"If my Harry couldn't get his supply of Mr. Atom's liniment," said a farmer's wife, "I don't think his poor muscles would make it through the winter."

"How long has he been using it?" I asked.

"Let's see," she said thoughtfully, "about fifty years." Some referred to him as "Doc" or "Professor." Now it was clear why he never expanded his enterprise with salesmen or advertisements—it would have deprived him of human contact.

On this Saturday night at Zern's, his lights were hooked up, the squawky sound system going, the crowd milling through in midway fashion. Families came closer to peruse the displays of weather-worn photos and clippings. On the lecture platform a remarkable transformation took place: there he was king.

"Hey, Atom . . . saw you on television," a young fan called out.

"What's My Line?," "Merv Griffin," radio, tv. He had been on twice in the past two weeks, but without Leah, the plaudits meant nothing to him. He thanked his admirers with a forced smile.

"Hey, Atom . . . there was a half page about you in the Philadelphia *Inquirer*." News stories had been circulated by United Press International.

He began his talk—health, clean living—berating the young men before him for smoking and throwing their health away. He began talking about sex and segued into observations on the Almighty.

"*I don't believe in God!* Belief is based upon superstition and ignorance of the facts. A man believes because he does not know. *I know there is a God!* Mere belief is no longer necessary. Knowledge sets man free.

"How do I know there is a God? Have I ever seen him? No. Ever shaken hands with him? No. But I am *alive*. I breathe. Have I ever seen the air I breathe? No. *But it sustains my life!*"

I sat under the umbrella at the table, taking orders and making change while he lectured. In demonstrating the potency of his liniment, he opened a bottle and poured some into his hand for those in the front row to take a whiff. Most sniffed and turned away, blinking from the acrid vapors. "It will give temporary relief for muscle and rheumatic pains."

A man of advanced years came toward me out of the crowd. He was dressed in farmer's coveralls and walking slowly with a cane. His face revealed his pain, his gnarled hands were half paralyzed.

"Son, I got bad rheumatoid arthritis. Do you think that stuff would work temporary for me?"

I waited for the Atom to pause, then spoke quietly so as not to disturb him. "Joe, is your liniment good for rheumatoid arthritis?" He nodded with finality, and resumed his discourse.

I opened a bottle and poured some into the man's hand. "Pat this on. If it works for you, come back and buy some." He dabbed the red liquid on, thanked me, and limped away.

Fifteen minutes later he returned grinning from ear to ear. "Can I have three big bottles, please?" He read the label intently. "It works, works." He quickly opened and closed the hand which moments before had appeared atrophied. He limped off with the bottles of Atom's Superior Liniment clinking in his coveralls.

"Hey, Joe," I called. "Your liniment just helped a fellow with bad arthritis." He looked at me with confusion. "If it didn't work . . . why would I sell it?"

On the bright afternoon of Father's Day, June 15th, I found myself with the entire Greenstein family who had gathered in Valley Stream, Long Island, to honor the patriarch. He sat in a lawn chair on the grass as family members approached with best wishes of the day. A grandchild came forward with an offering. "For you, Zayde [Grandpa]." A photographic book of Israel. He nodded a reserved thank you. One by one he received his family, from among his eight surviving children, twenty-three grandchildren, twenty-four great-grandchildren, and one great-great-grandchild.

He turned to me. "Would you believe two little people like me and my wife made all these people?"

The next week "Slim the Hammer Man" called long distance from Pennsylvania.

"Ed, have you seen Atom?"

"No, why?"

"I'm kinda worried about him. He took a fall today."

Joe, still preoccupied with the loss of Leah, had finished a lecture and was about to step down from his platform three feet off the ground. He tripped on a length of chain and pitched forward, falling head and shoulder down on the asphalt. The horrified audience was ready for an ambulance or a hearse. The Atom quickly jumped up, as if attacked from behind. The bystanders rushed to him.

"Atom, are you all right?"

"Sure, I'm all right." He brushed off his pants. "That's some step." He winked and continued about his business. As usual, and on time, the next lecture commenced with him railing against the evils of white bread.

I called him in Brooklyn that night.

"Atom, how do you feel?"

There was a momentary silence at the other end.

"Usually with my hands," he quipped.

He had no idea why I had called.

I stayed with Joe for a weekend at Zern's, sleeping in the truck. Early Sunday morning before the market opened, he set out his 2-inch pine board with the thirty sheets of tin atop it, roughly 3 inches of material. He wound up and brought his right hand down . . . whump! The twenty-penny nail in his hand smacked through the metal and into the wood, but did not pierce it completely. He examined it with concern.

He tried again, the spike a flash in his hand. Still, by a scant ⅛ inch it failed to penetrate.

"I'm sorry." He said nothing more, and retired to the truck. He had done this twenty thousand times before and if he could not do it now, it was because his life-force was fading. He rattled around inside the truck. I knew he felt bad.

In early summer of 1976 I was backstage at Madison Square Garden. After three standing ovations the Atom retired to his Spartan communal dressing room: a table, a couple of benches, a sink. He sat for a moment before packing his strongman's gear.

The board he had used at Zern's was propped up in the corner. A curious glint caught my eye and I walked over. The sharp end of a spike shone under the fluorescent light.

Moments before, onstage, he had cleanly driven a 6-inch iron completely through the board and metal so that now only ¼ inch of the top protruded, not even enough for the hand to grip. The spike that had twice defeated him at Zern's was now rammed through as if by a pile driver.

I called his attention to the board. He confided a little secret to me. "I'm not young anymore, Eddie. People think I'm something because of what I can do. They don't know . . . I'm holding back. If I don't it will shorten my life."

Russell C. Turiak

"This time you didn't hold back."
The ancient strongman nodded.

I sat at his kitchen table and studied the old man. The world had not dealt too kindly with Superman. Here he was, in his eighties, alone, with little money, his fame eclipsed. Perhaps the childhood dream of every American was misplaced. "If only I were Superman . . ." Variations on this Superman concept lay at the foundation of movies, adventure novels, and television series. Joe Greenstein could have been big business, but he lacked the cunning.

Instead he strove for the outer perimeters of physical and spiritual existence, and having reached for the stars without his feet on the ground, he paid the price.

Without being asked, the old man answered my ruminations.

"The Almighty must have put us here for a higher purpose than to accumulate wealth. How many meals can a man eat at one time, how many suits can he wear, how many beds can he sleep in? I figured I might as well do what gave me satisfaction."

I sat at the kitchen table with Superman. He riffled through his Yiddish poems.

A few weeks later he had sold the old house on East Ninety-sixth Street, and as planned had moved in with his daughter Esther and her husband Max in Brooklyn's Mill Basin section. They had prepared a private room for him, with desk and small library, radio and tv, and double-exposure windows. But nothing anyone did could please him.

"What's the matter, Joe?"

"I'm lost."

Two years after Leah died he was still inconsolable. He took his private hell with him, pacing the floor as if caged. All the paradoxes of his personality surfaced. He could drive straight through to Florida without missing a turn, but he got hopelessly lost in Queens.

"How could you get lost going to Florida? It's one way. In Queens you have to turn. . . ."

When asked why he couldn't meet a nice woman at the Senior Citizen's Center, he wrung his hands. "I can't. I don't know how to talk to women. I haven't got the guts."

In sixty-three years there had been only one; he was "Padlock Joe" after all. While he was desperate for female companionship, his standards excluded most of the females in the Western world.

"I have not squandered my life. I do not involve myself with idleness. I do not play cards or watch much television. I will not flit my time away, because that's what life is—time. I have always used mine to good advantage."

"Joe, what does that have to do with meeting a nice lady?"

"I don't waste my time," he explained, "and I don't associate with those who do. I don't like women who complain, gossip, play cards, or dance. I need a quiet, refined lady with a bright mind and a good heart. Someone like my Leah."

He called me several weeks later with the excitement of a schoolboy. "Eddie, I've met a woman at the center. A highly intelligent, cultured, and wonderful woman. I want you to meet her."

"What's her name?"

"Mollie Cohen."

On July 13, 1976, I was the first to arrive at the rabbi's home, a cheery place overlooking the water of Brooklyn's Brighton Beach. A half hour later, the doorbell rang and Joe arrived accompanied by his bride-to-be and several of his sons and daughters. I hadn't seen him for a few weeks and he embraced me. I stared at him: it wasn't the same man. He looked twenty years younger. His face was tanned, the beard neatly trimmed; the brow was relaxed and had lost its furrows. He was dressed in a new tailored suit with boutonniere.

Mollie Cohen had returned his old spirit to him; she had done so sympathetically, having lost her own husband some time before. She was a head taller than Joe, a refined and dignified woman who related not at all to his being the famous strongman. Her tastes ran more to operatic music and charity work than to iron men and the *Guinness Book of World Records*.

I brought several copies of the bicentennial issue of *Strength and Health,* the premier physical-culture magazine. In an article about great strongmen of the twentieth century, the last two columns were dedicated entirely to the Mighty Atom. He showed the magazine to his bride-to-be who examined it with curiosity.

"Is that really you, Joseph?"

"Sure, it's me!"

He looked at her for a long moment, then kissed her bashfully. "You've captured the tiger!" he said.

He at eighty-three, she at seventy-one, they were married in a simple touching ceremony nine days after the nation celebrated its bicentennial. There were twenty-five friends and relatives present,

Joe and Mollie Cohen, married in 1976. E. Spielman

blue-haired smiling ladies from the Senior Citizen's Center of Midwood; they took instamatic pictures and dabbed at their eyes.

Mollie was above all his pal. Though a very proper woman, she agreed to sleep in the old truck at Zern's—part of the honeymoon package.

The wedding reception was a joyous and exhilarating affair, held at a Brooklyn Kosher Chinese restaurant. Friends and relatives raised their glasses in a toast to the new couple, to the renewal of life. Joe's son Mike asked facetiously, "Pop, am I going to have a new brother or sister next year?"

"If you don't"—the Atom winked at him—"it won't be because I haven't tried."

TESTIMONIALS

By the mid-1970s the Mighty Atom had been generally accepted as the last remaining link with a bygone Golden Age of Strongmen, the man who was to human strength what Houdini had been to escapology, a patriarch of health and fitness whose dietary theories had come of age. Writings about the Atom consistently appeared in health, physical culture, and athletic journals, Willoughby's definitive volume, *The Super Athletes*, the *1976 Guinness Book of World Records,* Ripley's *Believe It or Not,* and newspapers and periodicals worldwide. Although he didn't have a press agent, the Atom's assemblage of scrapbooks weighed thirty-five pounds.

For all of this, the Atom remained an enigmatic figure whose importance had not been projected by the media or grasped by the public on the grass-roots level. Much of his television and print exposure represented him as little more than a human curio. His own view of the media was somewhat jaded. "When the Messiah comes, he's going to need a press agent," Greenstein said.

In his quest to share his knowledge and abilities, however, he had been successful. He had influenced and inspired several generations of Americans. A record of his peers and progeny reads like a "Who's Who" of strength athletes and physical culturists.

With traffic at its heaviest, a chauffeur-driven car stopped all traffic at the spot where the Atom was lecturing. Out of the back of the car

> jumped a well-dressed man shouting at the top of his lungs, "*God bless that man. He saved my life. Buy everything he sells!*" With that the gentleman jumped back into his car and off he went, allowing the traffic to proceed.

So went an account in *Strength and Health* magazine of December, 1962, in which Vic Boff reported an incident of the 1930s which he witnessed at New York's Union Square. Boff is an athlete, author (*Physically Perfect, Powerfully Strong*), lecturer, Doctor of Chiropractic, inventor, health-food-industry pioneer, one of the world's foremost winter bathers. President of The Iceberg Club of New York, Vic Boff finds peace in icy ocean water cold enough to stop a man's heart. He wrote many articles on Greenstein's amazing life after having observed him for decades.

Boff stated flatly, "In terms of strength into old age, Joseph L. Greenstein is the greatest of recorded history.

"There have been, and will continue to be great strongmen throughout the ages. But nowhere else in athletic annals is there a man who, in his eighties can perform with the same ease and perfection the variety of feats that he accomplished while at his height in his twenties.

"To my knowledge, there is no record of any other man who can, at such an age, perform the best feats of his youth on a daily basis. Greenstein appears to have broken the age barrier. He is truly a wonder of the world."

Randall Bassett is the New York-based author of *Zen Karate*, creator of the Zen Karate System of self-defense, instructor in survival techniques and mind control. He observed:

"If ever the building that houses my karate studio should catch fire, the one object I would grab as I ran out the door is a curious chunk of twisted metal—the remnant of a size No. 5 horseshoe that was ripped in two by the Mighty Atom in 1934. How often I have marveled that this same man, who was on the wrong side of forty at the time he made this break, is still making his living as a professional strongman in his eighties. There is something invincible about Atom, and for me it's symbolized by the odd-shaped piece

E. Allen Becker

of metal that sits in my studio. When visitors examine it, they seldom seem to be able to relate to it on anything less than a quasi-mythic level. Most of them conclude that Atom has a "secret" of some kind that allows him to do what no ordinary man could ever hope to do. . . .

"Feats that inspire wonder also tend to inspire fantasies . . . fantasies that get in the way of genuine insight. Certainly this was true in the case of Houdini. On more than one occasion he tried to talk seriously about the preparations that made him a nonpareil of stage illusion and escapology. No one really seemed to want to understand; they preferred their illusions. In the end, Houdini gave them the mythic figure they craved. He didn't have to preserve his "secrets"; the public's appetite for fantasy would ensure that they would never understand anything more than suited his purpose.

"In understanding the methods of such men as Houdini and the Mighty Atom, in my opinion, two factors stand out above all others and prove most instructive. The first of these factors had to do with a seeming deficiency. The personalities of Houdini and Atom do not encompass something more . . . so much as something *less*. A piece of their psyche seems to be missing in the sense that neither individual appears capable, or perhaps willing, to accept the concept of personal limitation. Certainly the greatest part of Joe Greenstein's talent consists of his liberation from self-limiting conceptions.

"The feats of men like Houdini and Atom would never have come into being without their also having had complete mastery of what I call *professional patience.* This is the second factor, the mechanism that moves them from the plane of the ideal to the actual. There is no better definition of professional patience than that given by Houdini when once asked what magic was. 'Magic is practice,' he answered.

"The very essence of such know-how consists of the capacity to endure those periods when there seems to be no growth in one's strengths and/or skills. Anyone can keep going when he's obtaining continuous positive feedback; authentic self-power largely consists of the trained capacity to do without such feedback. Those who master the skill almost always are thought of as wonder-workers.

"Naturally, there are countless things in this world that are impossible, even for the Atom. But he doesn't know this! And, of course, this is his genius. Men like Houdini and Atom decondition themselves from such psychic obstructions. They succeed in this deconditioning process because they perceive reality as being infinitely plastic. For them there are no obstacles for which nature has not provided levers sufficient to allow passage. If they do not always find these levers, it is merely a technicality . . . not enough time."

Joe Greenstein never once sought to publicize himself as what he was, in the author's firm belief, head to toe and pound for pound, the nearest thing to a superhuman that the world had ever produced. Never before on record had there been a man whose every bodily part exceeded the normal; Joseph L. Greenstein had every qualification for the title Superman, everything but a red cape and a birth certificate from the planet Krypton.

But the Mighty Atom had failed utterly in one crucial area, namely, self-promotion. In the *1976 Guinness Book of World*

Records, page 458, there is a credit for "World's Strongest Bite" and a photo of him cleaving a tire chain with his teeth.

"Joe," I asked, "you could easily have another half dozen or so world records, if you only bothered to show your documentation."

"The hell with it. I've got enough records. There's more important things to do. I'm writing a poem for my Mollie."

There were those who suggested that Joe Greenstein was a national resource, and wondered why no government agency had deposited him in a laboratory for study. In lieu of scientific analysis it would be safe to conclude that the Mighty Atom's immense internal and external power was an amalgam of unbroken decades of conditioning, a basically vegetarian raw-food diet, sound genetics, incentive borne of necessity, developed powers of concentration, self-hypnosis, and unconquerable will. These and spiritual/religious conviction created a Mighty Atom, considered by many of his peers to be the most remarkable strongman of all time.

"In the late 1920s I bought a picture of the Mighty Atom for twenty-five cents, and I still have it to this day. That's what an impression he made upon me." This encomium was offered by John Grimek, now editor of *Strength and Health* and *Muscular Development* magazines.

"He introduced me to strongmanship," said Grimek, who became the Babe Ruth of bodybuilding, a man who had developed such a perfect physique that he won every competition he entered. Because of John Grimek came the rule that a Mr. America title could be awarded only once to any individual. As a two-time Mr. America, Mr. Universe, Mr. U.S.A., Best Built Man of the Century, and Olympic weight lifter, the only other man to so inspire American youth to physical culture was Eugen Sandow at the turn of the century.

Grimek observed: "The Mighty Mite was a combination of power that made him a giant among his contemporaries. No one was ever disappointed in any of the strength feats that the Atom demonstrated . . . they were genuine. I have known him now for almost a half century, and have tried to remain in touch with him over these years. He is a legend, the last of the great all-around strongmen."

Other kudos came from no less a personage than "The Father of Weight Lifting," who contributed more than any other man to its becoming a sport. Bob Hoffman developed America's first Olympic Weight Lifting Team in 1932. Strength and health authority, founder of the York Barbell Company, philanthropist, writer, publisher, and a member of the President's Advisory Council on Physical Fitness, Hoffman is no dilettante; in his heyday he could press 260 pounds overhead with one hand.

"The Mighty Atom has a longer history of strength feats than any man now living. . . . I think the tire chain I saw him bite through was the greatest feat of strength of its kind that I have ever seen." Hoffman regularly attested to and documented in print the legitimacy of the Atom's impossible feats. In the York Hall of Fame, York, Pennsylvania, he exhibits artifacts of Greenstein's career including the last spike that the Atom bit in half at seventy-two years of age.

Among those who offered honors was Dan Lurie, former "Most Muscular Man In America" titleholder, the "Sealtest Dan the Muscle Man" of CBS-TV's "Big Top." At a formal dinner under the auspices of Lurie, now publisher and entrepreneur, Joe Greenstein was the first man inducted into the newly formed World Body Building Guild Hall of Fame.

The sentiments on the bronze plaque were apt:

World Body Building Guild
Hall of Fame Award
To Joe Greenstein, The Mighty Atom
75 years of inspiration
And Unbelievable Feats of Strength
. . . May 4, 1975, New York City

The Atom had become a cult figure among athletes of varied pursuits. George Dillman, a Fifth Degree Black Belt Master of

Okinawan Kempo Karate, winner of 320 awards for martial arts, is the head of the Dillman Karate Institutes with branches in four states, and was the karate instructor to Muhammad Ali:

"He changed my life. His lessons in power of mind over matter were a revelation." Dillman made a habit of bringing his karate students to Joe Greenstein's lectures, "because he greatly enhanced their training. In karate, we use the full weight of our body rather than wasting most of it as would the average person. Mr. Greenstein has been instrumental in teaching this concept."

Dillman related one lesson that Greenstein offered to his students:

"Most people, if told to close their eyes and meditate, are unable to do so. Their minds wander. The Atom asked them to keep their eyes open, and imagine that their outstretched arms were fire hoses. They could see out of them, while directing an unstoppable force of water. This thought process was applied to either arm, legs, hands, feet, fingers, toes. He asked them to assume this mental state before initiating physical action of any part of the body, and to practice tuning it on and off.

"They returned to the school relaxed, imbued with self-confidence, and actually feeling the presence of their inner energy. The mental process that he instilled virtually doubled their ability to perform."

Dillman recalled, "I attended one of his Pennsylvania lectures, and was in a crowd of about 250 people. A big slab-shouldered farmer spoke up. 'Mr. Atom, I honestly suspect that some of the things I see you bend are fixed.' The Atom took the insult with a shrug. The farmer then held up a large steel horseshoe. 'Sir, I've tried to bend this; my sons, all powerful men, have tried to bend it. And you can see that all of us just got nowhere. Now, if you really are who you claim to be . . . I'd like to see *you* bend it.'

" 'Give it here,' the Atom said, accepting the object, and looking at it for a moment with amusement. 'I can and will do it because I don't use the strength of my body but the strength of my mind. To prove what I say is true, I will bend this while very relaxed, while singing a little song.'

"You could hear a pin drop in the place as he took it in his hands,

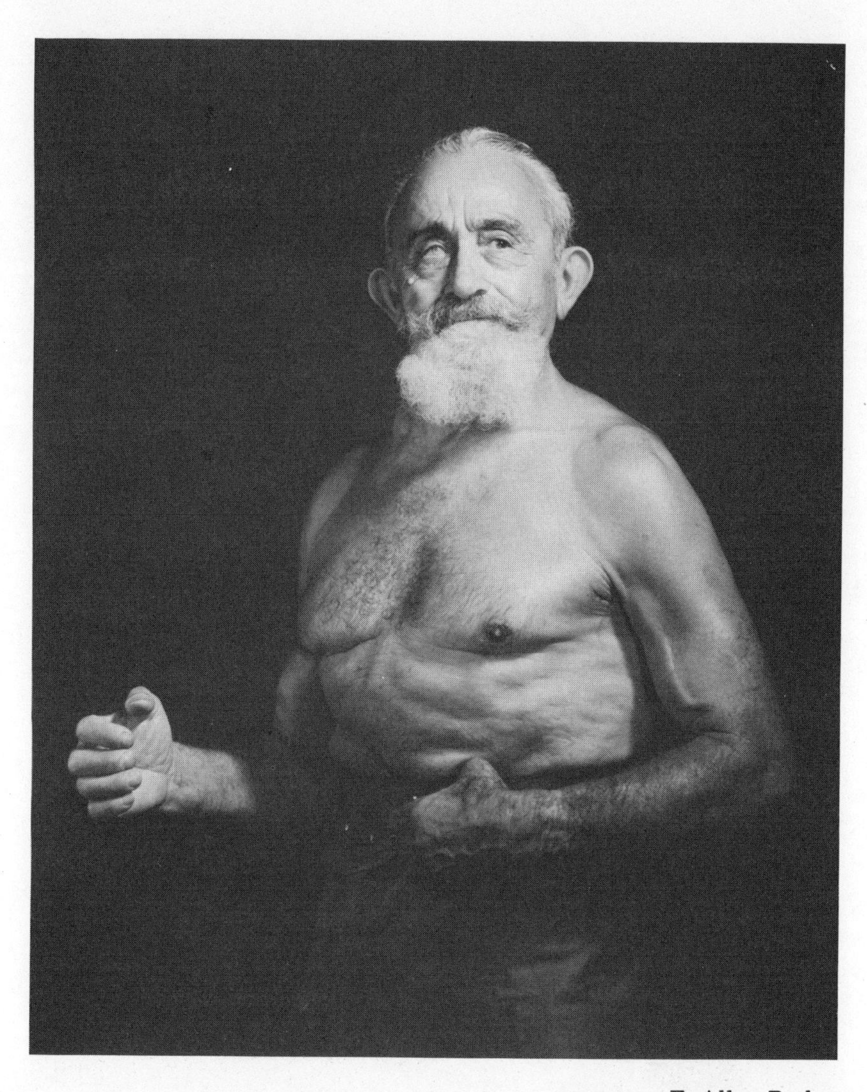

E. Allen Becker

sang his little song, and casually tore the horseshoe in half. 'You can have this back now,' he said, and returned it to the farmer who could hardly believe what he had just seen.

" 'Tell your sons to try using their minds sometime,' the Atom said, and continued his lecture."

As a token of his respect, Dillman was instrumental in honoring the Atom for his place as the oldest living Japan-trained Western practitioner of the martial arts. On his eightieth birthday, Joe Greenstein received an Honorary Sixth Degree Black Belt in Kempo Karate.

Of all Joe Greenstein's mementos, one that he perhaps treasured most had been given to him by the men of strength of Israel. As the tiny Jewish State was in the throes of its rebirth, he had done everything he could to aid the delivery, raising funds for the Irgun, lending his services to Zionist organizations, giving lectures and demonstrations. At a publicity event to attract volunteers for the reemerging Jewish State on February 10, 1948, the fifty-five-year-old Samson pulled with his hair a twenty-five-ton tractor trailer down Manhattan's East Broadway in front of the *Jewish Daily Forward* Newspaper Building.

He and Leah finally went to Israel in 1960, touring the country for

Tel Aviv, 1960: Joe next to his Israeli playbill

half a year, hosted by the Israel Defense Forces, visiting army camps and gymnasiums. A menorah lapel pin was presented to him in a meeting with Prime Minister David Ben-Gurion. Tel Aviv's Mograbi Theatre filled for a one-night-only exhibition by the sixty-seven-year-old strongman. Completing his performance, he was raised on to the shoulders of the crowd. The theater fell silent as he was presented with a silver cup, and its engraved sentiments read aloud:

> To Joe Greenstein, the humane and noble strongman, with love and admiration . . . In the name of all your admirers in the Holy Land.

"That honor touched me deeply because it was from Israel," he said. "It was so moving because, you see, I had to wait two thousand years to get it."

Among those to whom the Mighty Atom had become a legend, the question was often posed: "What is there in him that so inspires others?" The answer is as complex as the man, but Slim Farman answered it simply:

"I never had to prove to him that I was somebody. He told me I was."

THE LAST OF THE GREAT STRONGMEN

In July 1977 Joe moved his things to Mollie's cozy five rooms on the upper floor of a two-family house on Brooklyn's Avenue J. He ignored those few who thought this December marriage disrespectful to Leah's memory. No one could take her place, and he had never intended for anyone to try.

He and Mollie grasped all that time offered them. The Mighty Atom still walked spryly with an athlete's easy gait, he still had a small bit of money, he still drove his own car. It was a good marriage, each restoring that which had been lost in the other's life, rekindling a sense of purpose. Wherever they went, they held hands like a pair of teenagers. They made weekend jaunts to Zern's, continued with his benefit shows, attended awards dinners and body-building contests hosted by his friend Dan Lurie who often invited them to be judges, Mollie embarrassed by but glorying in this small notoriety. They traveled to Florida for a five-week vacation.

Shortly before his eighty-fourth birthday, Joe accepted an invitation to go to Denver to give several benefit performances and be honored as the founder of the Loong Kuen Pai Chuan Fa, a self-defense system of which his grandson Ron Rosen was chief instructor. Over family protests, he herded Mollie, Judith, and son-in-law Jack into the old white Chevy and took off for Denver, arriving three days later.

There, Mollie noticed that his color was poor, and suspected that he was in pain, though his sanguine manner covered it well. For the last of his performances, which had been reported by newspapers, local television, and United Press International, he readied a bar of steel which he routinely would have bent over his head at the bridge of his nose. But now the old strongman's knees buckled and he collapsed. "Again," he said, quickly regaining his feet. The bar gave way.

He drove home to family concern, insistence that he see doctors, take tests. They needn't have bothered. He knew what he had, and had known for some time: it was cancer. The last battle of his life had begun.

He did not brood, and conceded nothing to his illness. He took Mollie shopping, granted interviews, fought traffic tickets in night court. "The judge knew me!" He beamed. "He tore up the ticket."

They had been married for a year and two months. "We've done so much together," Mollie said. "I never dreamed I would have someone like Joseph. We lived every day. I have no regrets."

By September Joe lay in a Brooklyn hospital bed; the pain came intermittently, and for the moment he talked and joked and received visitors with his usual exuberance, though he had aged years in the last few days. I sat beside him.

"Joe," I said, knowing that flattery always seemed to cheer him, "my money's on you. The Mighty Atom's cheated death a hundred times. You're going to walk out of here."

He waited until Mollie was out of earshot. "Who do these doctors think they're kidding? Eddie, this time I don't walk away, I know I haven't got very long."

I held his hand. The time for pretense had ended. I met his gaze, unable to turn away, unable to shrink from this last insight into his character. "Are you afraid to die, Joe?" I asked.

"The Jewish way of death is very poetic," the rabbi said. "After his gift of life, a man leaves simply, just as he entered it, his body unaltered. The man is buried in a plain wooden box with no metal fittings, nothing to be left behind as his physical form returns to nourish the earth. And in that place of death"—he nodded with his prophetic smile—"in time, a tree will grow . . . *life.* No, I'm not afraid. I just don't want to suffer."

Another would have succumbed easily, but the Mighty Atom's heart and respiratory system was that of a young man. Only the dreaded cancer could cut down the old lion, and he knew he would go by inches. Long ago, after breaking three ribs during a vaudeville performance, he had taped them and the next day pulled a chain of trucks around the block without complaint. This same man who had never taken so much as an aspirin, now cried out for drugs to ease his pain.

And still he was the Mighty Atom. When a man in his room fell from bed, Joe sprang up, lifted him, and put him back, though moments before his own pain had immobilized him so that he could not stand long enough to go to the bathroom.

The Jewish New Year came in mid-September, and Joe was awake and lucid for the Rosh Hashanah service that was held in his room. All over the world with hope and prayer, his people were calling in the New Year of 5738 of the Hebrew calendar with a blast of the ram's horn. In a last act of defiance against the sickness that so terribly embraced him, Joe Greenstein raised the ram's horn to his lips and with his fading breath sounded the call, as if by sending the high shrill trumpet sound to the heavens he had announced his own imminent arrival.

On September 26, Slim and Shirley Farman drove five hours through the rain and arrived at the hospital where Joe lay. Mollie was beside him as she had been every minute since they met. The Atom's children and grandchildren filled the room, and stood grimly out in the corridor. Now the old strongman was in comatose sleep, his breathing coming in gasps. Slim Farman towered over the bed.

"He's full of drugs," Joe's daughter Esther said, "he probably won't know you."

The old man's eyes fluttered open. "Slim," he said barely audibly. The big man drew nearer to hear his words.

"Slim . . . ," the Atom said in a whisper, "give me a kiss." The giant kissed his forehead tenderly for a long moment, and stroked his hair.

"Joe," he whispered, "if I could take the pain, the dying for you, I would."

We stood in the hospital parking lot under a light rain.

"I'm not going to see him alive anymore, am I?" the big man asked.

"No, Slim," I said. "You're not."

He shook his head. "How will I ever get used to that?"

At ten A.M. on the gray morning of October 8, 1977, Joe Greenstein passed away at Brooklyn's Kingsbrook Hospital.

The sky was overcast as the funeral cortege wended its way across Queens' Springfield Boulevard. During the burial, as if on cue, the clouds parted and the sun warmed the place where the last of the great strongmen was laid to rest.

EPILOGUE

Hackensack, New Jersey
July 23, 1978

He stood before the crowd, his face a mask of the intense concentration required of his craft. Emblazoned on the chest of his black tunic was a gold Star of David. The only sound to be heard was that of man against steel. And when spirit and muscle had exploded the last chain, the crowd rose with a clamor, in recognition of the proof that dreams can be made real, that man has no bonds which cannot be undone if only he wills with enough courage. The strongman was Slim "The Hammer Man" Farman. The life-force continues on.

E. Allen Becker

The Mighty Atom

THE GREATEST ATTRACTION OF ALL TIMES
Will Appear on the Stage of this Theatre

Tues., Wed. & Thurs., April 9—10—11

THREE TIMES DAILY

The GREAT ATOM is the name of the strongest lightweight man that ever lived in the centuries past... He is without doubt the Samson and the marvel of the age. His feats of strength thrill and amaze. The late Breitbart who was considered the world's giant, was 6ft. 2 inches high, and weighed 265 lbs., while the Great Atom's height is only 5ft. 4 inches, and weights 145 lbs.
THE GREAT ATOM EXHIBITS THE FOLLOWING FEATS OF STRENGTH:

1—He drives a 20 penny nail through a 2½ inch pine board with the palm of his hand.

2—He shapes a horseshoe out of a piece of raw steel, twelve inches long, one inch in diameter, by bending it across his thigh.

3—He wraps across his arm a piece of steel 8 feet long, one-half inch thick, and an inch and a quarter wide, as if it were tin. It would take a blacksmith one hour to temper, heat and bend it, whereas the Great Atom does it in less than a minute.

4—He bends a raw piece of steel, 18 inches long and 1 inch in diameter, across the bridge of his nose.

5—He takes a piece of steel, a half inch thick by one inch wide and makes a corkscrew of it.

6—Like Samson, the Great Atom possesses twice as much strength in his hair as in his arms, body and teeth. He bends a piece of steel 2 feet long and three-eighths of an inch thick, by an inch and a quarter wide, on the strength of his hair.

7—He pulls a piano weighing 2000 pounds with six people on it across the stage by the strength of his hair.

8—He pulls three 7-passenger automobiles filled with people by attaching a chain to a comb which is wrapped in his six-inch long hair, and tows the three cars. To start the three automobiles requires a power equal to twenty-one thousand pounds.

9—Besides his strength, the endurance of the Great Atom is absolutely supreme and unique. He lays down with his bare back on a bed of one thousand needle pointed spikes. A platform weighing 450 pounds is placed upon his chest. Two husky men weighing 500 pounds sit on it and two girls dance upon the platform, making a total of 1200 pounds. It requires twice as much strength to balance the platform while the girls are dancing on it, than to hold the enormous weight.

Tues., Wed. & Thurs., April 9—10—11

THRILLING — SENSATIONAL

The Mighty Atom

WILL POSITIVELY APPEAR IN PERSON

The Strongest Man in the World in his amazing feats of STRENGTH

Do Not Miss This Treat!

THE MIGHTY ATOM will give an exhibition of his strength on Tuesday, April 9th at 12 o'clock Noon at the corner of DELANCEY and NORFOLK STREETS.

ברוך הבא

דער מאָדערנער שמשון הגיבור איז דאָ

דער שטאַרקער אטאָם

אַפּאָלאָ טהעאַטער

126 קלינטאָן סטריט

דינסטאָג—מיטוואָך—דאָנערשטאָג

אַפּריל 9—10—11

דער וואונדער פון יעצטיגער צייט

דער מענש וועלכער ווערט בעוואונדערט פון אַלע פעלקער און בעגריסט פון די עכסטע מענער פון לאַנד און בעקומט דעם כבוד און דעם שליסעל פון יעדער שטאָט וואו ער קומט און ער פערדינט דאָס. זינט דעם פערלוסט פון אונזער בעליבטען און מעכטיגען העלד זיסע ברייטבאַרד, ז. ל., האָט נאָך קיין מענש ניט אומשטאַנד גיוועז צו בעווייזען די וואונדער און גבורה וואָס דיזער נייער העלד איז אומשטאַנד צו בעווייזען; ער איז אונקאַמפּערעבעל און דער סוקסעספולער זיגער איבער דיא מעכטיגסטע מענער פון לאַנד — עס איז אייער פערלוסט ווען איהר פעלט עס צו זעהן. מיר מעגען שטאָלץ זיין מיט אונזער וואונדער מענש. זיין געוויכט איז 148 פונט, איז 35 יאָהר אַלט, ער קומט פון סובאַלק, פּוילען נערוועז פון אייזען און מוסקולען פון שטאָל.

קומט אונד איבערצייגט זיך דעם אמת.

PER 3 GIORNI SOLI	*TEATRO* **APOLLO** 126 CLINTON STREET	PER 3 GIORNI SOLI

Martedi Mercoledi Giovedi

Aprile 9—10—11

Masista De Cabria

Sa Ra Pesente

Il Grande Atom e il nome del forte peso leggiero cheha, esistito nel il cesimo centennario il Sansone e la Maravighia la sua forza inmenza vi dara molta emozione e divertmento lesposizione de la grande forza del GRANDE ATOM sara esibita all ore iz di giorno al can tome di Norfolk and Delancey Streets MARTE di APRILE 9 non vi di

menticate di vederlo